THE SUPPLY CHAIN
GAME CHANGERS

THE SUPPLY CHAIN GAME CHANGERS

Applications and Best Practices That Are Shaping the Future of Supply Chain Management

MIKE BURNETTE
PAUL DITTMANN, PH.D.
TED STANK, PH.D.

Publisher: Paul Boger
Editor-in-Chief: Amy Neidlinger
Executive Editor: Jeanne Glasser Levine
Consulting Editor: Chad W. Autry
Cover Designer: Chuti Prasertsith
Managing Editor: Kristy Hart
Project Editor: Andy Beaster
Copy Editor: Cenveo Publisher Services
Proofreader: Cenveo Publisher Services
Indexer: Cenveo Publisher Services
Compositor: Cenveo Publisher Services
Manufacturing Buyer: Dan Uhrig

© 2016 by Michael H. Burnette, J. Paul Dittmann, and Theodore (Ted) Stank
Published by Pearson Education, Inc.
Old Tappan, New Jersey 07675

For information about buying this title in bulk quantities, or for special sales opportunities (which may include electronic versions; custom cover designs; and content particular to your business, training goals, marketing focus, or branding interests), please contact our corporate sales department at corpsales@pearsoned.com or (800) 382-3419.

For government sales inquiries, please contact governmentsales@pearsoned.com.

For questions about sales outside the U.S., please contact international@pearsoned.com.

Company and product names mentioned herein are the trademarks or registered trademarks of their respective owners.

Printed in the United States of America

First Printing October 2015

ISBN-10: 0-13-409378-X
ISBN-13: 978-0-13-409378-9

Pearson Education Ltd.
Pearson Education Australia Pty, Ltd.
Pearson Education Singapore, Pte. Ltd.
Pearson Education Asia, Ltd.
Pearson Education Canada, Ltd.
Pearson Educación de Mexico, S.A. de C.V.
Pearson Education—Japan
Pearson Education Malaysia, Pte. Ltd.
Library of Congress Control Number: 2015946710

Contents

Foreword

The University of Tennessee is one of the highest-rated Supply Chain Management schools in North America. The University is host of the Global Supply Chain Institute (GSCI). This Institute links the SCM undergraduate program, graduate programs, and executive education programs with our business sponsors and networks. The GSCI sponsors the largest university supply chain forum in North America with over 60 corporate sponsors and 2 annual Supply Chain Forums (over 200 supply chain professionals in attendance). The GSCI provides our business partners with supply chain audit capabilities, corporate SCM training programs, and consulting. The GSCI can be reached through the University of Tennessee or by searching "Global Supply Chain Institute" on Facebook, LinkedIn, or Twitter.

Acknowledgments

We would like to acknowledge our GSCI sponsors (over 60 corporations representing $1.7 trillion annual revenue) and our GSCI advisory board (40+ senior, executive supply chain officers) for their proactive support, including networking, benchmarking, coaching, financial, and project partnerships. These leading companies are dedicated to delivering game-changing supply chain improvements.

About the Authors

Michael H. Burnette is the Associate Director of the Global Supply Chain Institute (GSCI) at the University of Tennessee. Burnette comes to the University of Tennessee after a 33-year career as a Supply Chain executive at Procter & Gamble. Most recently, he was the P&G Global Supply Chain leader for Skin Care (owning 2+ billion dollar Olay brand) and P&G Global Supply Chain Leader for Hair Care (owning 4 billion dollar Pantene and Herbal Essence brands). His supply chain leadership and expertise include supply strategy/design, manufacturing, logistics, innovation, acquisitions, and human resources.

Currently, Burnette teaches supply chain courses at the University of Tennessee, manages multiple GSCI projects (including the GSCI white papers), is a supply chain consultant, and is an event speaker.

J. Paul Dittmann is Executive Director of the Global Supply Chain Institute at the University of Tennessee. In addition, he teaches supply chain courses in the business school, and lectures in the executive education programs.

Dr. Dittmann comes to the University of Tennessee after a 32-year career in industry. He has held positions such as Vice President, Logistics for North America, and Vice President Global Logistics Systems, and most recently served as Vice President, Supply Chain Strategy, Projects, and Systems for the Whirlpool Corporation.

Dr. Dittmann also manages many of the special projects done for companies, including the supply chain audits. In managing these audits and projects, he has consulted or done executive education for numerous firms and organizations, such as Walgreens, Pfizer, Walmart, UPS, Boise, Tyco, Inmar, Honeywell, Genuine Parts, Cintas, Cummins, Cooper Tire, United Smokeless Tobacco, Rhodia, Radio Systems Corp., Johnson & Johnson, Estee Lauder, the United States Army, the Marine Corps, Michelin, Brunswick, Nissan, Lockheed

Martin, RaceTrac Petroleum, GAF Corporation, Edison Schools, OfficeMax, Sony, Keller Group, GlaxoSmithKline, Cooper Tire, Lowes, Navistar, GAF Corporation, Fiskars, Edison Schools, and the United States Air Force.

He has also taught numerous public seminars in the areas of lean manufacturing, global business, and supply chain excellence, and has spoken at many conferences on these and other topics. He has been a certified instructor for the Project Management Institute.

Dr. Dittmann co-authored a recent Harvard Business Review article, *Are You the Weakest Link in Your Supply Chain*, and also co-authored the book, *The New Supply Chain Agenda*, published by Harvard Business Publishing. Another book, *Supply Chain Transformation: Building and Implementing an Integrated Supply Chain Strategy*, was published by McGraw Hill in August 2012.

He is on the Board of Directors of Kenco Group, a member of the University of Missouri Industrial Engineering Hall of Fame, was selected as a Rainmaker by DC Velocity Magazine, and was designated one of the Top Ten Supply Chain Thought Leaders in 2013.

Dr. Stank is the Harry and Vivienne Bruce Chair of Excellence in Business in the Department of Marketing and Supply Chain Management at the University of Tennessee. He assumed the Bruce Chair following nearly 6 years in administration as Department Head for Marketing and Logistics, Associate Dean for Academic Programs, and Associate Dean of the Center for Executive Education. Prior to arriving at University of Tennessee, he served at Michigan State University (1997–2003), Iowa State University (1995–1997), and the University of Texas at El Paso (1994–1995). He holds a Ph.D. in Marketing and Distribution from the University of Georgia, an M.A. in Business Administration from Webster University, and a B.S. from the United States Naval Academy.

Dr. Stank's business background includes sales and marketing experience as an employee of Abbott Laboratories Diagnostic Division. He also served as an officer in the United States Navy prior to his industry and academic experience. He has performed consulting and executive education services for over 50 organizations, including: Dell, EDS, Kellogg's, IBM, Lowe's, Norfolk Southern, OfficeMax, Pepsi, Siemens, Sony, Textron, Walgreens, Walmart, Whirlpool, and the US Marine Corps. He is the current Chairman of the Board of Directors of the Council of Supply Chain Management Professionals (CSCMP). He also serves as Educational Advisor for the *Journal of Business Logistics* and serves on the editorial review board of *Journal of Operations Management* and *International Journal of Physical Distribution and Logistics Management*.

Dr. Stank's research focuses on the strategic implications and performance benefits associated with logistics and supply chain management best practices. He is author of more than 90 articles in academic and professional journals, including *Journal of Business Logistics*, *Journal of Operations Management*, *Management Science*, *Journal of Retailing*, *Supply Chain Management Review*, and *Journal of the Academy of Marketing Science*. He is also co-author of the books *Global Supply Chains: Evaluating Regions on an EPIC Framework (Economy, Politics, Infrastructure, and Competence)* and *21st Century Logistics: Making Supply Chain Integration a Reality*, and co-editor of *Handbook of Global Supply Chain Management*. He has received numerous awards for his research and teaching, and was named a Logistics Rainmaker by *DC Velocity* magazine.

Introduction

At the University of Tennessee, we do a great deal of research on leading-edge supply chain concepts that run the gamut from future supply chain megatrends to logistics operations. One publication even ranked us number one in the world for supply chain academic research output.[1]

We have learned much about modern supply chains and even more about how to manage them. But after numerous conversations with industry partners, we concluded that we needed to do a better job reaching the practitioner community. We were not adequately communicating our findings to the supply chain professionals who were doing the real work, or to the leaders who were struggling with the daily challenges of supply chain operations.

Supply chain professionals are very busy people. They are continuously engulfed in a maelstrom of events ranging from angry customers to damaged shipments, and they all must be addressed immediately. They know they need to keep developing professionally, but have precious little time to do so—and even less time to read documents written by academics, for academics. What we had to do as supply chain researchers, then, was publicize our game-changing findings as accessible, highly substantive material.

In 2012, we launched the *Supply Chain Game Changers* series of white papers to do just that. We have written six of these white papers thus far, with topics ranging from global supply chain management to distribution-center best practices. These papers have enjoyed great success in the professional community, so we thought it was time to put a bow around that work and publish it in a book that supply chain professionals could use as a single reference on a variety of topics that

---

[1]M. Maloni, C.R. Carter, and L. Kaufmann. *International Journal of Physical Distribution and Logistics Management*, 42, no. 1, 83-101, 2012.

we will describe later in this introduction. Each white paper represents a chapter in this book. As the series of white papers continues to grow, we will periodically publish additional sets of workbooks for the many dedicated practitioners of supply chain management.

One of the fundamental concepts of this book is that of game-changing trends in supply chain management. These trends are significantly impacting how leaders shape their supply chains today. Our research allowed us to group these trends into four major areas as shown in Figure 1.

The six topics in this book fit nicely into this framework. In Figure 2, you can see the first topic, "Game-Changing Trends," at the core of the diagram. The "Global Supply Chains" topic is next in an inner circle touching all of the subjects. Then the four remaining

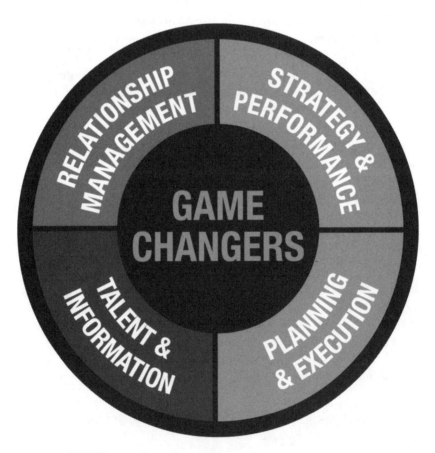

Figure 1 GSCI SC game changers model

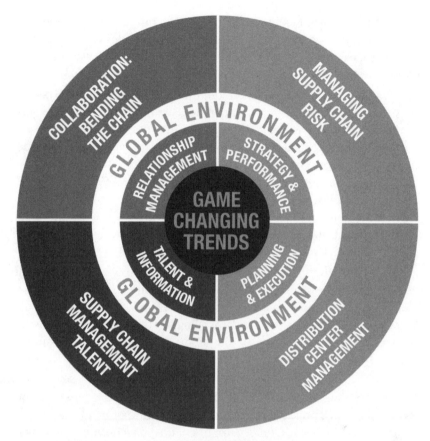

Figure 2 GSCI SC game changers model link to this book

chapters reside on the outside: "Managing Risk in the Supply Chain," "Collaboration: Bending the Chain," "Supply Chain Talent Management," and "Distribution-Center Management."

Chapter 1 starts at the core, with a description of 10 game-changing trends in supply chain management. Fresh ideas about game-changing trends arise constantly and are highly dynamic. In this chapter, we define game-changing trends as those trends that are extremely impactful on a firm's economic profit and shareholder value while also being very difficult to successfully address. In addition, we included an addendum that takes a long look ahead and considers how these trends may play out by 2025.

Our interactions with hundreds of companies helped us develop this list of 10 game-changers. Not only do we identify the supply chain

trends that you will face in modern industry, but also we provide recommendations for how to make progress toward desired end states.
We also use plenty of real-world examples along the way to make
these concepts more applicable to typical business practice.

Chapter 1 also includes an addendum. While the changes occurring in supply chain thought and practice during the first 15 years
of the 21st century were significant, many experts predict that even
more substantial change will occur over the *next* decade. In fact, the
supply chain world of 2025 promises to look very different from the
one today. This addendum extrapolates on the state of the original
10 game-changing trends in order to predict the evolution of supply
chain management over the coming decade. In addition, it identifies
five new trends that are likely to be influential in driving the changing
supply chain between 2015 and 2025.

Chapter 2 then moves to the topic that arguably encompasses all
of the others: the global supply chain environment. Most firms have
global suppliers and/or global customers. Supply chain professionals
know that they somehow need to manage this complex worldwide
network in order to provide better service to customers while simultaneously delivering lower costs and inventory levels.

In this chapter, we provide a series of best-practice recommendations to help meet these daunting challenges. These are based on the
EPIC (economy, politics, infrastructure, and competence) framework
from the recent book, *Global Supply Chains: Evaluating Regions on
an EPIC Framework.*[2] In this chapter, we also break down the best
practices for supply chain network design in the global environment,
as well as best practices for managing complex global supply chains.
Chapter 2 is intended to provide valuable tools for supply chain leaders to design and manage a *winning global supply chain.*

Chapter 3 explores risk in the global supply chain. Over the last
decade, many companies have faced extreme supply chain challenges
that have stretched their capabilities to the breaking point. Both the
preponderance of natural disasters and huge economic swings have
caused challenges across the supply chain. These challenges are certainly not going away.

---

[2]M. Srinivasan, T. Stank, P. Dornier, and K. Petersen, *Global Supply Chains:
Evaluating Regions on an EPIC Framework—Economy, Politics, Infrastructure, and
Competence* (New York, NY: McGraw-Hill Professional, 2014).

Supply chains, which once functioned almost on autopilot, face many dangers today in both global and domestic markets. Due to its global nature and systemic impact on the firm's financial performance, the supply chain arguably faces more risk than other areas of the company. Risk is a fact of life for any supply chain, whether it is dealing with quality and safety challenges, supply shortages, legal issues, security problems, regulatory and environmental compliance, weather and natural disasters, or terrorism. This chapter discusses best practices for identifying, prioritizing, and mitigating risk.

Chapter 4 explores warehouse management best practices, a topic we refer to as the "ABCs of DCs." We felt that it was important to have a topic in this book dealing with supply chain and logistics execution. Logistics professionals who operate distribution centers (DCs) have a tough job. Management constantly challenges these leaders to cut cost, which basically means doing more with less. With cost cutting as their primary focus, they also need to make sure that customer responsiveness does not suffer, and in fact improves. This calls for a highly advanced management skill set. Chapter 4 covers a wide range of DC management best practices, from picking/receiving to people to sustainability.

Chapter 5 delves into the very challenging field of cross-functional integration. In the supply chain audits that we have conducted over the years, we do many interviews with managers and executives—more than 800 in the past 5 years alone. We always end each interview with a "wish list" question. What topped these wish lists most often was the desire for all functions in a company to work together toward and align perfectly for a common purpose. It is no exaggeration to say that the professionals we interview pine for an environment where the functional silo walls have come down. They intuitively know that these disconnects are the real reasons things are not improving faster.

Chapter 5 specifically focuses on one of the greatest potential threats to integration. Ironically, the biggest threat can be found between two interactions traditionally thought of as belonging to the supply chain function: the interaction between purchasing and manufacturing, and the interaction between manufacturing and logistics. Our research suggests that a major gap often exists between purchasing and logistics, resulting in the destruction of value. This lends credence to the infamous quote, "We have met the enemy and he is us!" Chapter 5 discusses the best practices for bridging these divides.

Chapter 6 takes a deep dive into talent in the global supply chain. Ten to fifteen years ago, the supply chain leader in most companies held the *Vice President of Logistics* title. This was a largely functional role that relied on technical proficiency in discrete areas: knowledge of shipping routes, understanding of warehousing equipment, familiarity with distribution-center locations and footprints, and a solid grasp of freight rates and fuel costs. The Vice President of Logistics reported to the Chief Operating Officer or Chief Financial Officer, had few prospects of advancing further, and had no exposure to the executive committee.

Thanks to game-changing global trends, the way companies need to think of the modern supply chain executive has changed dramatically. In Chapter 6, we describe the professional skills leading supply chain professionals need today, as well as the best practices for managing talent. In doing so, we introduce a five-stage framework of talent management:

1. Analyze needs
2. Find talent
3. Recruit talent
4. Develop talent
5. Retain talent

The conclusion to the book takes all of this material and synthesizes it. We identified the top 10 actions that we strongly believe you need to take to create a world-class supply chain. In fact, we included a self-evaluation tool to help you assess exactly where you stand today. This chapter is intended to bring it all together for you, and help you prepare an action plan to *change your supply chain game*.

We hope that you enjoy the journey through this supply chain game-changing material. You might want to jump around, going first to the topics in which you have the most interest. We have compiled this book in a way that allows you to do just that. Or you may find it more useful to read from beginning to end, in the order we agreed was most logical. However you use this book, we hope you will gain insights along the way that can truly transform your supply chain.

1

Game-Changing Trends in Supply Chain Management

First Annual Report by the Supply Chain Management Faculty at the University of Tennessee

Ted Stank, Ph.D., Chad Autry, Ph.D., John Bell, Ph.D.,
David Gligor, Ph.D., Ken Petersen, Ph.D.,
Paul Dittmann, Ph.D., Mark Moon, Ph.D.,
Wendy Tate, Ph.D., and Randy Bradley, Ph.D.

Introduction

Every month a new article or conference lecture seems to present a fresh idea about the game-changing trends faced by supply chain professionals. Consulting companies, academics, and even individual companies have their own opinions about and definitions of supply chain megatrends. Often these megatrend lists do not match; instead, they reflect the backgrounds and experiences of the people who compile them.

For this chapter, we define game-changing trends as *those trends that meet the two basic criteria of being extremely impactful on a firm's economic profit and shareholder value, as well as very difficult to implement successfully.*

The 10 game-changing trends discussed in this chapter have their foundations in a 2000 landmark study[1] by Don Bowersox, David

---

[1] D. Bowersox, D. Closs, and T. Stank, "Ten Mega Trends That Will Revolutionize Supply Chain Logistics." *Journal of Business Logistics* 21, no. 2 (2010): 1–15.

Closs, and Ted Stank at Michigan State University. That study and the updated version conducted by University of Tennessee (UT) faculty in 2012 incorporated responses from supply chain professionals across a wide range of industries (retailers, manufacturers, and service providers, large and small in size). However, that does not mean the material in this chapter is dated—far from it. Based on our annual interactions with hundreds of companies (the largest industry network in the academic world), we believe that those trends still serve as a foundation. But the supply chain world has drastically changed over the past 13 years. *In this chapter, we will discuss the game-changing supply chain trends that you face today, and we will suggest how to make progress toward the desired end state.* We will also use plenty of examples along the way.

When a student asked Albert Einstein if this year's physics exam questions were the same as last year's questions, he responded, "Yes, but unfortunately for you the answers are very different." Our game-changing trends are like those questions. We still see some of the same supply chain trends, but the real-world responses to them are dramatically more sophisticated.

The University of Tennessee is ranked number one in the world in supply chain management research.[2] Eight members of our supply chain management faculty collaborated to identify 10 game-changing trends that form the basis for this document. The 10 trends align perfectly with the research of these faculty members. We tapped our wide-ranging experience with industry leaders through our global forums, executive education, and consulting.

This chapter provides a brief synopsis of today's leading thinking about 10 game-changing trends in supply chain:

1. Customer service to customer relationship management
2. Adversarial relationships to collaborative relationships
3. Incremental change to a transformational agile strategy
4. Functional focus to process integration
5. Absolute value for the firm to relative value for customers
6. Forecasting to endcasting (demand management)
7. Training to knowledge-based learning
8. Vertical integration to virtual integration

---

[2]*International Journal of Physical Distribution and Logistics Management* (2012).

9. Information hoarding to information sharing and visibility

10. Managerial accounting to value-based management

In the last two months of 2012, over 160 supply chain professionals were surveyed to assess these 10 trends in industry, as well as how those trends have changed. While they will be expanded in the body of the chapter, here are some highlights of that survey:

- Firms have made significant and, in some cases, *surprising* progress in the last decade and in each of the 10 areas.

- Some companies have achieved top levels of performance in individual categories, although no firm has excelled in all categories. Thirteen years ago, no firms reported a top level of performance in *any* category.

- There are laggards in each category as well. These firms appear to be fighting the same battles in the same way.

- Respondents feel that most progress has been made in customer relationships and cross-functional integration. Firms seem to be better focusing on their customers outside the firm and shoring up the emphasis on cross-functional processes inside the firm.

- Talent management clearly emerges as the linchpin required for advancement in all areas.

The remainder of this chapter is divided into 10 sections, one for each trend. We used the survey information as one input but also included the leading-edge thinking from all of our research and industry contacts. The evolution of these 10 trends is visualized in Figure 1.1.

Customer Service to Customer Relationship Management

By Ted Stank, Ph.D., Harry J. and Vivienne R. Bruce Chair of Excellence in Business

Why Is This a Game-Changing Trend?

Customer relationship management remains a game changer for all companies because it requires customer prioritization to maximize revenue and profitability by targeting limited resources. Few would disagree that this needs to be done; yet we find that few companies actually create customer-focused, differentiated supply chain plans.

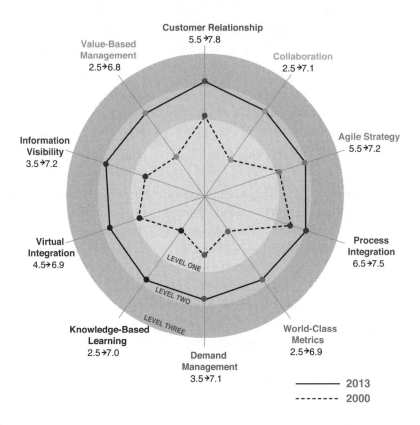

Figure 1.1 Significant work remains on game-changing capability

They require tough choices that are sometimes unpopular and often engulfed in politics. But a few leading firms have truly embraced this game-changing trend that is gathering momentum.

One medium-sized retailer, a member of our Supply Chain Forum, survived the Great Recession by focusing on critical customers. They confirm that this segmented customer focus was the crucial element in a strategy that allowed them to take out 48% of their inventory while still improving on-shelf availability from 96% to 98.8%.

Progress Made but More Needed

We have come a long way as a profession by changing our focus from merely providing excellent customer service to actually managing relationships with targeted customer segments. What will it take to reach the next level in managing customer relationships? Before answering that, let us reflect on the progress that has been made (see Figure 1.2).

In the 2000 landmark study referenced in the introduction, the authors noted that the customer service to relationship management trend was one of the most advanced game-changing trends in supply chain. On a scale of 1 to 10 (10 being total adoption and 1 representing no meaningful acceptance), it had achieved an average score of 5.5 on level of maturity achieved. The prediction was that, by 2010, organizations would be operating close to the total adoption level, equivalent to a score of 10. The research supporting this chapter suggests that industry has continued to improve on this trend, although most organizations still have not achieved total maturity. The average among the firms responding to the 2012 survey was 7.85, with a range (at one standard deviation) of 5.88 to 9.82. The important takeaway is that while some leading firms have pushed close to total adoption,

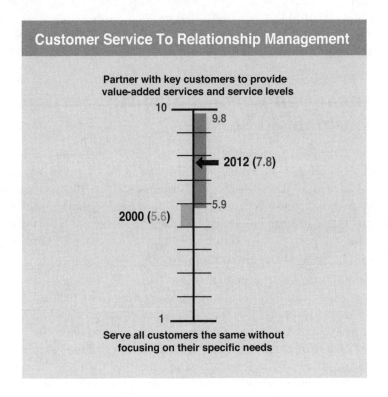

Figure 1.2 The shift to targeted customer service has started

a significant number of organizations may be considered laggards in adopting this game-changing trend.

Many Firms Still Operate With a Traditional Service Mindset

In the past, the traditional approach to customer service was standardizing service for all customers. The original study showed that many firms operated from a traditional service mindset even as the industry had progressed toward managing each customer, or customer segment, as a unique service relationship. It should be noted that by understanding true "Cost-to-Serve" of a client/product segment, a company can more effectively allocate/charge for the services provided/consumed.

Companies recognized that supply chain performance on dimensions such as in-stock availability and on-time delivery was critical to the buying process. Thus, they developed standard operational capabilities to facilitate standard and uniform levels of service delivery to all customers. With that focus, achieving internal performance goals and reaching targeted operating standards were indicators of success (e.g., on-time delivery, shrinkage levels, number of customer complaints).

From a High Level of Standard Service to Customized Service

Improving operational effectiveness for all customers is highly desirable, and may be considered a building block toward gaining competitive advantage. However, business success derived from a focus on standard levels of service effectiveness is usually short-term at best. The managerial tools and techniques utilized are typically easy to imitate, so performance differences gained from such programs are difficult to sustain in a tough competitive environment.

In addition, a traditional customer focus that results in standard levels of service for every customer risks creating value that misses the mark for all customers. For example, delivery features such as shortened cycle times and exact point-in-time delivery may be the prime drivers of performance for some customers. But others may prefer average or lower levels of delivery support rather than pay the cost of such high-level operations. Service delivery should reflect the unique requirements of each customer.

Customizing service offerings that already demonstrate high levels of operational excellence provides an opportunity for a supplier to become an integral part of a customer's business. A supply chain based on close customer relationships has the greatest potential to generate unique solutions that combine elements of timeliness, availability, and consistency to exactly match desired values at prices customers are willing to pay.

Creating close customer relationships helps a firm identify the long-term requirements, expectations, and preferences of current and/or potential customers. It also enables the firm to develop operational configurations that deliver tailored supply chain delivery with optimal profitability. A company may attain a competitive edge through close customer relationships. Those relationships enable them to become more proactive, anticipate customer expectations, and measure the extent to which customers' needs are satisfied.

Multiple Supply Chain Configurations Are Required

To generate a unique delivery value tailored to specific customers or customer segments, firms will have to create multiple supply chain configurations. This tailored approach to customer service requires a huge investment to establish close relationships and deliver customized value. That investment is a major challenge.

No firm possesses sufficient resources to successfully meet the exact needs of every potential customer or market segment. Firms are forced to focus resources on select customers and segments that represent the best exchange opportunity for the type of value created by the organization. The firm must decide where to compete and where not to compete based on the fit between its strengths and its customers' needs.

The success of the tailored service relationship depends, therefore, upon managers' understanding of their strengths in comparison to the differing needs and desires of each customer or customer segment. Once the specific needs of each customer or segment are understood, the segments must be prioritized based upon their strategic importance and their potential for economic profit.

While this game-changing trend was first envisioned for manufacturers serving business customers, it may become the standard operating procedure for retailers as they increasingly move into multichannel supply chains. In a multichannel retail case, the prioritization would be focused on the preferred channel rather than on the specific customer.

The Fallacy of Being All Things to All Customers

Highly volatile competitive environments often pressure firms to abandon the segmented service relationship approach and focus on trying to be all things to all customers. During such times, senior management becomes fixated on the mass market and loses sight of the costs and asset commitments needed to deliver unique value to customers. The result is that both financial and human resources are dedicated to customers or customer segments that are unlikely to generate a profit. Although such short-term strategies can augment cash flow in the near future, they are inevitably damaging to profitability and earnings.

Other competitive scenarios can create monopolistic-like competition in customer markets and lead to the opposite problem: organizations with minimal viable competition focusing unduly on supply chain efficiencies across the entire customer base to the detriment of creating customer value, thereby eroding service levels in the pursuit of low-cost/low-asset operations. Neither excessive focus on customer effectiveness nor excessive focus on supply chain efficiency presents an optimal strategic situation. Each de-emphasizes the importance of understanding the needs of each major customer or customer segment and then matching that with the appropriate service configuration. The goal is to create customer value commensurate with the economic profit potential of the relationship.

Challenges of Prioritizing Service Levels

A couple of factors are at work here, namely, *service levels* and *service level offerings*. There are additional differences that depend on the industry. In all cases, there is a need to prioritize by customer. Firms face numerous challenges as they seek to focus their resources more on prioritizing service to customers of choice. First, marketing and sales organizations are typically reluctant to cast any paying customer in a role of "less important." This often has less to do with desire and more to do with a lack of accurate and timely activity-based costs tracked to specific customers to enable a reasonable analysis of customer profitability.

If the actual profitability of customers, both current and future, is not clear, prioritizing service could be fatal. Still, some firms have begun allocating costs to customers on activity drivers that are more meaningful than the standard usage of percent of sales. Percent of total orders placed or percent of total volume per order could more meaningfully approximate a true cost-to-serve.

A second challenge to prioritization is a lack of useful implementation tools. For example, the order-management systems in most companies cannot "hold" inventory so that a top customer is prioritized over less important customers during stockouts. Also, top customers often demand shorter order cycle times, and inventory may be listed as unavailable (as it was already promised to another customer) by the time the customer of choice places an order. Some progress in this area has been made by assigning customer teams that may manually override system designations. Recent advances in inventory optimization software now let manufacturers automate the process of managing different service levels for different customers within a specific distribution center.

Our research partners at Ernst & Young have adopted an extension of the relationship management idea. The concept—called *service stratification*—applies not only to service differentiation by customer segment, but also to the concurrent differentiation by product group and product offerings. Stratification of service levels and offerings on the basis of customer and product value groupings requires organizations to establish a holistic perspective of both customer and product value. Leading companies have developed a robust governance structure to drive consistent execution of segmentation strategies. The operational impact of stratified service policies can be demonstrated at the execution, tactical, and strategic levels of a company's supply chain.

Successful execution of service stratification requires multiple steps:

1. *Define business policies and rules*: The first step is to define the categories within the service stratification. What constitutes an A, B, or C product/customer, based on the company's business model? For each category, a set of business rules should be developed that indicate how each category in the matrix should be treated.

2. *Integrate business policies*: Next, the business has to integrate the business rules into daily processes across the firm at the strategic, tactical, and executive levels.

3. *Develop policies that are automatic/systematic to drive the most value*: The final step, which promotes continuous improvement, is to start to automate policies and procedures within the processes that will create high value if automated.

Software exists to model and optimize inventory, taking into account different service levels for different customers or channels (same item, same distribution center, different service levels). Optimizing as a group instead of individually as a specific customer class significantly reduces overall stocks by location.

We conclude this chapter with a case study that provides an example of a firm that has established close relationships with select customers. This partnership has allowed the firm to provide differentiated levels of service that enhance the value provided to customers.

Example:

- A consumer packaged goods (CPG) firm recently started focusing on choice retail customers and partnering with key customers to develop a more strategic view about how the business should progress. The company is now involved in joint business planning with its top customers. This is a radical shift from the previous business model that focused on cost reduction. This approach ". . . made us order takers rather than demand creators," said a senior manager in one of our interviews. More importantly, the new approach has been coupled with the need to balance supply chain capacity with the desired demand that is being created. This change has come from the highest leadership in the organization and is a significant cultural change for the organization's managers and employees. To embed the new approach into the organizational culture, a balanced scorecard has been adapted to include common metrics across all functional areas and support the new customer focus on relevant value. The company is in mid-transition, but has a strong framework that is guiding decisions throughout. Financial results are beginning to suggest the success of their new business approach.

Adversarial to Collaborative Relationships

By Chad Autry, Ph.D., Taylor Professor of Supply Chain Management

Why Is This a Game-Changing Trend?

Developing collaborative relationships with suppliers and customers is a game changer because so few firms really accomplish true win-win partnerships. But the few that do have experienced dramatic and even breakthrough improvements in product availability, cash flow, cost, and shareholder value. Supply chain professionals have been talking about collaboration for years, but unfortunately, as one executive lamented, "When all is said and done, there has been more said than done." Companies can achieve game-changing competitive advantage in this area by accomplishing what their competitors have failed to do.

For example, in the book *The New Supply Chain Agenda*, we discuss in detail a win-win collaborative relationship between OfficeMax and their supplier, Avery Dennison.[3] The collaboration results were spectacular:

- In-stock fill rates rose from about 90% to 99% plus.
- Lead times fell by 60%.
- Forecast accuracy improved by 30%.
- Inventory turnover increased by 9%.

And beyond just the numbers, the two companies could focus on driving growth rather than confrontation and firefighting.

Encouraging Evidence of Collaboration

Outstanding examples of collaboration have finally begun to emerge. What will it take to get to the next level in collaborating with partners? Before answering that, and with Figure 1.3 as our guide, let us reflect for a moment on the progress made.

The 2000 study referenced in the introduction noted that many firm-to-firm supply chain relationships were problematic. The relationships were characterized by an overall lack of trust between parties, limited or sporadic information sharing across the organizational interface, and the pursuit of short-term, firm-specific benefits at the expense of partners' economic interests. To measure the effectiveness of a firm's collaboration, the original study used a 10-point scale. A score of 1 signified overt competition or hostility with a lack

---

[3]R. Slone, J. Mentzer, and J. Dittmann, *The New Supply Chain Agenda: The Five Steps That Drive Real Value* (Boston, MA: Harvard Business School Publishing Corporation, 2010).

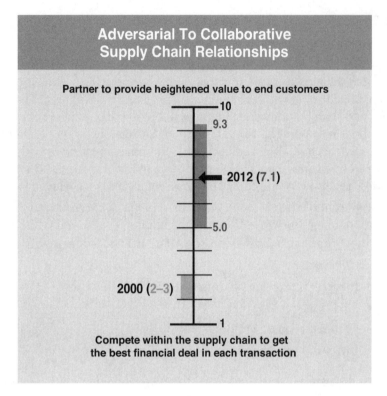

Figure 1.3 Collaboration is improving

of constructive relational behavior between the "collaborators." A score of 10 indicated full-scale partnering with focused activities and reciprocal accountability for reaching collaboration outcomes. At that time, overall supply chain collaboration across industries averaged between 2 and 3. Prospects for successful supply chain integration were bleak because the average relationship was adversarial and firms were more likely to offer lip service rather than actually practice collaboration. But the authors predicted that, by 2010, a typical collaboration across industries would score much higher. The current (2012) data collection confirms that prediction. Firms across industries range between 5 and 9.3; they average 7.1 on the same scale. This increase offers encouraging evidence that contemporary firms are using collaboration to reduce work duplication and redundancy, streamline workflows and processes, and communicate properly about shared objectives. Two companies working together can provide new value that cannot be accomplished by either company in isolation.

A Three-Part Best-Practice Model

Still, progress toward strategic collaboration within supply chain relationships has proven difficult for many companies, particularly if it requires changes to organizational culture and structure. If firms can improve on a limited set of identified behavioral drivers, collaborative behavior could potentially be leveraged further with significant impact.

Many of the issues that caused some collaborative planning, forecasting, and replenishment (CPFR) implementations to fail are still with us today. Future collaboration can successfully avoid these traditional challenges by using data analytics to improve service and grow revenue, to cut joint stocks to reduce capital investment and carrying costs, and to improve overall margins.

Our best-practice model indicates three initial drivers to facilitate a move toward better collaborative relationships. First, both parties need to address collaboration's potentially negative aspects. Procedures should be established early to resolve potential disputes, unforeseen issues, and potential dissolution of a dysfunctional or useless relationship. Second, to share risks and rewards, both parties must develop supporting organizational and interorganizational structures. This includes work rules, leadership roles, and guidelines that both parties will follow about how responsibilities, risks, and rewards will be shared. Third, mutual trust must be encouraged for strategic and operational integration. This includes establishing shared values and vision. In that shared vision, both must prioritize long-term viability to serve the end customer even when that requires a sacrifice of optimal short-term results. Considered together, these three groupings of relational aspects are critical to the success of collaborative ventures across businesses. Over the last decade, they helped move collaboration beyond that initial low score.

Supply Chain Collaboration—Additional Key Success Factors

As the discipline of supply chain management has evolved over the past decade, three more specific drivers have emerged. Collaboration commitment, goal congruency, and integrated information sharing have been identified as critically important. Each of these success factors is explained in the following with current examples to depict best practices. By focusing on these drivers, businesses can develop superior connections with key partners. That ultimately leads to a competitive advantage with greater efficiencies and market effectiveness.

Collaboration Commitment

For collaboration between two supply chain firms to succeed, mutual trust remains essential. However, simply trusting the partner to act as expected is risky. External forces, internal interests, or a lack of commitment to the venture may disrupt the best collaborative intentions. A senior manager may be unwilling to underwrite a project even though the day-to-day participants are heavily invested. Inadvertently or purposefully, firms can violate the agreement.

In the mid-2000s, Procter & Gamble (P&G) initiated an automatic inventory replenishment project with Kmart stores to reduce the total cost for selling baby products. At the time, those products represented a critical source of profitability for P&G. P&G invested significantly in the program. However, Kmart managers never really bought into the project. They believed it was too tactical to yield major transformation, and so after an only marginally successful early experience, their attention waned. With Kmart's low expectations, the collaborative venture underperformed. P&G learned from the experience and was able to establish successful inventory replenishment arrangements with Walmart later.

In a nutshell, firms can no longer simply create procedures for disagreements; they must also define social or financial costs for violating agreements. These costs disrupt relationship-damaging behavior and reduce failure risk. When both sides see each other as having some "skin in the game," each can more readily justify resource expenditure toward the common goal. If P&G and Kmart had developed a mutual understanding of goals early in the collaboration period and then established performance metrics that adequately incentivized each party to participate, the replenishment venture would have had a better chance of succeeding.

Goal Congruency

Rather than simply establish shared values and visions, collaborating parties need to share desired outcomes as well. Establishing common goals appears to be a relatively straightforward task, but the devil is in the details. Sometimes one side assumes performance specifics that are not covered by the formal agreement, or the collaborating parties establish common goals but prioritize them differently. One party's goals may change as the collaboration develops, but the other may wish to stick with the original agreement until the venture is complete. In each of these instances, the parties disagree about the desired collaboration outcomes. That lack of alignment can lead managers to adopt a "what is in it for me" attitude toward the collaboration. Then the desired collective outcomes suffer.

On the other hand, clear goal and outcome congruency can lead to mutual benefits. When DuPont Chemical sought to become a more sustainable company in the late 2000s, its collaborative venture with Exel Logistics began with several meetings designed to create goal congruency and define the "mutual win" space. At these meetings, senior executives from both companies openly discussed the best way to significantly reduce DuPont's 81 million pound annual landfill volume. With a deep understanding of DuPont's ambitions, Exel crafted solutions such as composting, reutilization, and waste-to-energy. These reduced landfill volume more than 20% and reduced CO_2 emissions by nearly a million metric tons within 5 years.

Integrated Information Sharing

Rather than simply create generally supportive organizational and interorganizational structures, collaborating firms need to integrate technology to share real-time information. The value of firm-to-firm information sharing in facilitating supply chain integration is well established. As technology has become easier to acquire and develop, it has become a necessary but insufficient criterion for supply chain success. Accessing technology does not necessarily create value. Partners must analyze the data for opportunities and threats to shared goals; then they can pre-emptively solve problems. An information-sharing culture must be developed to support the technology. A venture is likely to underperform if users on both (all) sides are unwilling to enable openness and transparency via the selected technological solution.

If an information-sharing culture can be developed, collaboration capabilities and benefits are likely to increase. For example, when a key supplier for road construction firms recognized significantly greater demand for its materials along with pressure to execute on-time, low-cost deliveries, it sought a collaborative solution. The supplier partnered with a leading technology provider and a long-haul carrier. Together they customized a transportation-management system to minimize road miles (and therefore fuel expenditures) while also minimizing how many underpasses and narrow-laned and/or urban roads would be used. The solution worked because the customer firms were willing to estimate job times and locations well in advance. They also shared their forecasts for material requirements. Collectively, they were able to create a mutually beneficial venture.

Summary

Firms are establishing fewer adversarial and more collaborative relationships than they were a decade ago. The climate has changed, but the adversarial overtones that have characterized many supply chain collaborations have not entirely disappeared. Increasingly, however, firms are realizing the benefits of employing the tools used to enhance collaboration, including commitment, goal congruency, and integrated information sharing. The long-term performance outcomes emerging from such collaborative relationships are gradually overcoming the lingering hesitancy toward their use.

Incremental Change to a Transformational Agile Strategy

By John Bell, Ph.D., The University of Tennessee, and David Gligor, Ph.D., Assistant Professor, University of Mississippi

Why Is This a Game-Changing Trend?

Our research shows that only 16% of firms have a documented, multi-year supply chain strategy at all, let alone an agile, transformational strategy. Yet the supply chain is the heartbeat of the firm and the prime driver of economic profit and shareholder value. Firms must have a strategy if they hope to achieve a game-changing supply chain.

In the *Harvard Business Review* article *"Leading a Supply Chain Turnaround,"* the authors discussed the story of how Whirlpool implemented an agile, transformational strategy in its supply chain.[4] The strategy served as the foundation for a true supply chain transformation and led to the following breakthrough results:

- Historically low inventory level, down almost $100 million
- Record high service levels
- Total cost down over $20 million (ignoring inflation)
- Major customer satisfaction improvement, with typical customer survey responses being

---

[4]R. Slone, "Leading a Supply Chain Turnaround," *Harvard Business Review* 82, no. 10 (2004, October), 114-21, 158.

o "Whirlpool is most improved."

o "Whirlpool is easiest to do business with."

o "Whirlpool is most progressive."

o "Whirlpool is good now, but more importantly, they are consistently good."

Most Firms Lack an Agile Transformational Strategy

In the landmark 2000 study, researchers optimistically predicted that firms would move from an environment dependent on experience and incremental change to a transformational strategy capable of adapting to unprecedented competitive changes. And our survey results—displayed in Figure 1.4—do reveal some movement in that direction.

Back in 2000, many felt that uncertainty and continuous change in the competitive—and rapidly emerging—global marketplace would force firms to rely dramatically less on historical processes and strategies. They predicted that 21st century firms would need to continually

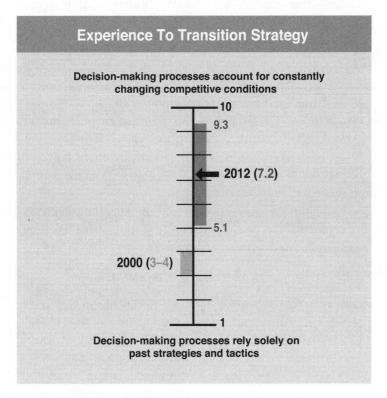

Figure 1.4 SC strategy quality is improving but we must be more agile

reinvent processes with decision-making capabilities to quickly and flexibly adapt to unexpected environmental changes. Agility would be the order of the day. Firms would have to learn how to manage in "uncharted waters," to recognize that previously successful actions may not be effective in modern conditions.

Three Activities Needed to Be More Adaptive and Agile

In 2000, researchers noted that to develop more adaptive and agile transformational strategies, firms would need to take three actions:

1. Identify and document an expanded total landed cost-to-serve as a framework based on multiple scenarios and not limited to historical solutions.
2. Develop business analytics and modeling skills for situations outside the norms of current business.
3. Build better expertise in the application of decision support and IT tools to identify patterns, assess potential performance, and manage newly developed processes.

Progress has been made in each of these dimensions, but two important concepts could help firms move more rapidly to a transformational strategy. First, the development of "agile" supply chain strategies requires the proactive and adaptive mechanisms needed to respond to unprecedented marketplace changes. Second, building and executing an agile strategy requires a management team with the right mix of talent and decision-making skill sets to navigate a firm through uncertain times.

The remainder of this section focuses on several concepts that will help firms move further toward achieving a score of 10 in this megatrend.

The Essence of Agile Strategies

Agile strategies recognize that individual businesses no longer compete as solely autonomous entities. Rather, businesses compete as *supply chains*. To achieve a competitive advantage in the rapidly changing business environment, firms must coordinate operations with suppliers and customers to achieve a level of agility beyond that of competitors. Supply chain members must be able to rapidly align collective capabilities to respond to changes in demand and supply.

Agility is a broad concept that can be defined as *the firm's ability to quickly adjust tactics and operations within its supply chain to respond or adapt to changes, opportunities, or threats in its environment*. New research being conducted at the University of Tennessee (UT) indicates that supply chain agility possesses five distinct dimensions: alertness, accessibility, decisiveness, swiftness, and flexibility. Many of these dimensions are analogous with sports medicine's concept of agility as applied to world-class athletes and the military's perspective on agility as applied to fighter pilots. The following sections describe each of the five dimensions of supply chain agility in detail.

Alertness

Alertness is the first dimension of supply chain agility, and it can be described as *the firm's ability to quickly detect changes, opportunities, and threats*. This dimension suggests that firms must recognize changes before they can respond to them. Within sports science, research has shown that a player's ability to execute tasks is dependent upon factors such as visual-scanning techniques, visual-scanning speed, visual processing, perception, and anticipation. These factors are reflected in a player's on-field agility.

Elite performers differ from nonelite performers in their ability to anticipate their opponents' movements. In military science, researchers refer to the alertness capability as situational awareness, the perception of environmental elements with respect to time and space. Military forces in combat require early awareness of potential threats, and the faster the environmental changes are detected, the sooner a response can be deployed.

Similarly, agility in a supply chain setting requires that firms be sensitive to external markets (including competitors, customers, and suppliers) and their customers' changing requirements. A firm must be able to identify shifts in market trends, supplier capabilities, competitors' actions, and even government policy and regulations. Highly effective organizations capable of transition strategies remain alert to change and can successfully predict competitors' actions.

Accessibility

Accessibility emerged as the second dimension of firm supply chain agility and is described as *the ability to access relevant data*. UT research suggests that once a change is detected, a firm must also be able to access relevant data to decide how to respond. Supply chain–wide information access is a key requirement for supply chain agility. This implies that agile supply chains must be virtual—that is, they must be information-based rather than inventory-based. Supply chain members must share real-time demand, inventory, and production information to build a more transitional strategy.

In military science, the ability to orient the combat unit to the situation and determine potential courses of action is critical. In the military, these capabilities reside in the intelligence and communications functions. Similarly, for competitive firms the creation of a virtually connected supply chain allows individual supply chain members to access and communicate relevant data in real time and then make informed decisions about how to respond to detected environmental changes.

Decisiveness

Decisiveness is the third dimension of supply chain agility and can be described as using the available information *to resolutely make decisions*. High performance in sports is ultimately determined by effective decision-making skills. Offensive players who demonstrate proficient agility employ superior decision-making skills in response to the movements of their opponents; as the complexity of the task increases, decision-making skills become even more important.

Similarly, firms must create processes to sort through available information and make resolute decisions on how to respond to supply chain changes. They cannot rely on only historical experience, which may no longer be relevant. Processes such as Sales and Operations Planning (S&OP) help provide a forum for resolute decision making that utilizes the best available information in demand and supply markets.

The alertness, accessibility, and decisiveness dimensions of agility emerge through coordination and planning processes that enable firms to determine appropriate responses to opportunities or threats. Agile coordination and planning processes are necessary, but they are not everything required for supply chain agility. A firm must also be able to act on agile decisions. Swiftness and flexibility are the capabilities that firms use to implement agile decisions.

Swiftness

The fourth dimension of agility, swiftness, is defined as *the ability to quickly implement decisions*. Sports and military science identify swiftness as an essential component of agility. In both, the speed of movement and a change of direction speed (or action) are required to respond agilely.

Similarly, to achieve the desired level of supply chain agility, firms must develop the ability to complete an activity as fast as possible. For example, within manufacturing, the ability to carry out tasks and operations in the shortest possible time has been considered a necessary condition for agility.

Flexibility

Flexibility is defined as *the ability to modify the range of tactics and operations to the extent needed.* In a sports context, an athlete's mobility of joints (i.e., flexibility) controls the range of quick adjustments they can perform. The type of directional change (agility) depends on the flexibility of the specific body parts involved in the exercise. If an athlete exceeds his range of flexibility when attempting to perform a maneuver, injury is likely to occur.

Sports science research indicates that agile performance can be improved through flexibility training. Military science research also recognizes that built-in flexibility is needed for agile military response. In the same way, a firm's supply chain operates within a specific range, and the firm's supply chain agility (i.e., adjustment of tactics and operations) will be constrained by that range. For example, the firm's supply chain cannot quickly produce more items than its fixed manufacturing capacity allows. Therefore, a firm's response to changes depends on the flexibility of its supply chain tactics and operations.

Agility in Practice

Together, the five distinct dimensions of agility allow firms to rapidly respond to a volatile and ever-changing marketplace. According to research sponsor Ernst & Young, many of today's leading firms are attempting to achieve these agile capabilities by adopting innovative practices that include benchmarking, purchasing intellectual property rights, or even acquiring smaller start-up firms to extend their capabilities.

Similarly, research sponsor Terra Technology has indicated that agile capabilities are no longer just about using internal decision support systems, as indicated in the original research study. Today, firms need the ability to automatically collect, analyze, and make daily decisions based on large volumes of external data. Demand-sensing technologies and decision algorithms with artificial intelligence capabilities now allow firms to adapt in an agile manner.

According to Terra Technology, this means having seamless automatic software capabilities that can retrieve retail point-of-sale information on a daily basis. In this manner, agile responses include knowing what is going on *and* knowing what to do about it. Research and experience indicate that as firms move to more transition-oriented strategies, agile capabilities will be critical to achieve future marketplace advantages.

An Agile Workforce Is Critical

Today, building an experienced workforce that is trained only on historical standards and processes is not the right approach. Instead, recruiting and sustaining a labor force with keen decision-making skills appears to be critical for achieving an agile transition strategy that executes the five dimensions of agility. Firms will need managers with improved planning capabilities and state-of-the-art analytical skills to match an increasingly volatile supply chain environment.

According to Ernst & Young, competitive firms in today's dynamic business need a "people agenda" that builds the right mix of talents within their firm. Additionally, firms must understand how intellectual property and knowledge-based capabilities help them be more agile. According to E&Y, many of today's large, world-class corporations are studying and seeking to acquire the more agile decision-making skills possessed by smaller innovative businesses. The ability to rapidly rebuild structures and implement innovative solutions depends on the intellectual competencies possessed by a firm's leaders and workforce. Creating such capabilities can make a firm more agile and more competitive.

Summary

Over the last decade, firms have moved significantly toward agile transitional strategies. Now, they are much less dependent on experience-based strategies. However, by *understanding and improving their supply chain agility* dimensions, firms may continue to advance in this megatrend. Recruiting and developing talented employees with the decision-making skills needed in an agile environment remains critical for maturing in this area.

Functional Focus to Process Integration: Cross-Functional Integration—Purchasing Through Logistics

By Ken Petersen, Ph.D., Dean of the College of Business,
Boise State University

Why Is This a Game-Changing Trend?

Over the past 5 years, we have interviewed over 700 supply chain professionals during supply chain audits. Based on all of those interviews, we are absolutely confident that supply chain professionals recognize true cross-functional integration as a game-changing trend. In fact, when we ask the "wish list" question in our interviews, supply chain professionals do not ask for more resources but instead pine for a company where the silo walls have come down. They intuitively know that this lack of cross-functional integration is a huge issue holding them back. Oliver Wight International has documented the power of cross-functional integration in their work in integrated business planning (IBP). In our Supply Chain Forum, they shared the following results for the cross-functional collaboration's best performers (see Table 1.1).

Deep Cross-Functional Integration Is Still a Major Challenge

Most supply chain professionals would agree that cross-functional integration is an absolute requirement for supply chain maturity. Unfortunately, most would also acknowledge that they are well short of this goal. What will it take to get to the next level in cross-functional integration? Before answering that, let us reflect for a moment on the progress made.

In 2000, firms reported that they had achieved a level of 6 to 7 on a 10-point scale of functional orientation (1) to process integration (10). This can be seen in Figure 1.5. Today, firms are reporting that they have achieved an average level of 7.5, with a range of responses between 5.5 and 9.6. This score indicates that while much progress has been made, much work remains to be done.

Why Companies Need to Change

Companies must engage in this area for four key reasons:

1. Volatility is the new norm.

Table 1.1 Cross Functional, Business Integration has Improved But is Less Than 50% Effective

Cross-Functional Collaboration's Best Performers		
Area	Percent reporting improvement	Percent improvement
Forecast accuracy	43%	18–25%
Asset utilization	39%	
Customer satisfaction	39%	
Inventory reduction	37%	18–46%
Fill rate	34%	10–50%
Revenue increase	31%	10–15%
Working capital	30%	
Perfect order	30%	
ROA	30%	
Gross margin	29%	
Cost reduction	29%	30–45%

Note: Information presented by Oliver Wight International at the University of Tennessee Supply Chain Forum, November 13, 2009

2. Segmented supply chains are necessary to handle the differences (e.g., between how you manage high volume/low variability products versus low volume/high variability products, or small, highly profitable customers vs. large, not so profitable customers).

3. Agility and responsiveness are necessary to capture margins and/or profitable growth.

4. Erosion of traditional competitive advantages such as changing cost structures (low-cost manufacturing is no longer low cost).

Companies need to focus on the next generation of cross-functional integration. This will involve a move from the *traditional* view—sales/marketing plus supply chain (design-plan-buy-make-move) integration—to the *progressive* view—sales/marketing plus supply chain (design-plan-buy-make-move), finance, legal/regulatory, and order fulfillment/customer service. This will depend on:

• Cross-functional participation in decision making
• Shared information/data visibility and transparency

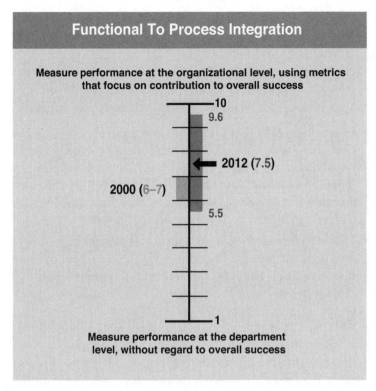

Figure 1.5 Overall functional collaboration improvement is limited

- Total cost/margin optimization versus functional cost optimization
- The ability to
  - o Assess tradeoffs holistically
  - o Respond to each value chain segment
  - o Manage profitability through the product and customer lifecycle

The Challenge: Business-Wide Cross-Functional Integration

Cross-functional integration is perhaps one of the most challenging opportunities for businesses today. Often, a business will consider upstream integration with its supply base or downstream integration with its customer base to be its most central integration problem. However, we believe that the largest integration problem is cross-functional and exists within the four walls of the business. In fact, this problem causes internal business functions to compete with each other, which reduces the ability to deliver customer value and perhaps even destroys customer and shareholder value. Cross-functional integration presents a difficult problem for a number of reasons. Some of these reasons include the firm's operating model, metrics, supply chain alignment, culture, and tools. Each of these enablers/barriers is discussed in the following:

- *Operating model*: Traditionally, firms have organized to deliver customer value by developing functionally specific organizational designs. From the perspective of overall business process improvement and customer value creation, this focus on the function has fostered excellence within function, but has not yielded excellence in the *connection* of functions.

- *Metrics alignment*: Often, metrics tend to drive *functional* performance but are disconnected from *cross-functional* performance. For example, the purchasing function may seek to reduce per unit raw material cost with longer lead times and larger lot sizes, while the manufacturing and planning functions seek to create more agile and responsive operations with smaller lot sizes and reduced finished goods inventory. Given the cross-functional nature of supply chain business processes, these functionally oriented metrics can drive functions to conflict and/or compete with one another.

- *Supply chain strategy alignment*: The supply chain strategy is often not aligned with a firm's core competencies or its competitive strategies. This results in a less-than-ideal collaboration within the supply chain functions as well as with other business functions. The supply chain strategy must enable corporate business strategy. For instance, the purchasing function may well be driven by a functional strategy that is neither informed nor understood by other functions within the business.

- *Culture*: Frequently, businesses have not invested in developing a deep culture of cross-functional collaboration. Of course, cross-functional business processes tend to be better supported by a culture of cross-functional integration in some areas. An example of one of these business processes is S&OP. However, the state of "cross-functional culture" in businesses is generally poor. Interestingly, when asked, most business leaders will state that their businesses are very mature at cross-functional integration. This misalignment between perception and reality hinders the development of a healthy cross-functional culture.

- *Tools*: The business tools for important processes simply do not support cross-functional integration and involvement. For example, many corporations run their supply chains from Excel spreadsheets rather than through integrated planning and execution systems. This lack of tools and information technology systems that facilitate collaboration results in inefficient manual processes. Valuable resources are spent to gather and crunch data rather than to analyze information, evaluate options, and make decisions.

In summary, we believe that cross-functional integration is a problem that exists across the business and all of its important functions. Furthermore, without healthy cross-functional collaboration, establishing an efficient end-to-end supply chain with trading partners (such as customers and suppliers) is much more difficult. To reiterate this point, business leaders must first create strong and robust cross-functional integration within their firm. Only then will their business have the opportunity to reach its integration potential with upstream suppliers and downstream customers.

An Unrecognized Problem: Purchasing and Logistics Integration

Next, we will consider cross-functional integration between the purchasing and logistics functions. This is a problem that has a tendency to fly under the radar in most firms. The lack of integration between two traditional supply chain functions is surprising and can be quite damaging to the firm. The focus of the cross-functional integration problem is often between operations/supply chain and sales/marketing. Yet the opportunity for integration within the supply chain area—and, in particular, between purchasing and logistics—is huge.

The Fallacy of Supply Chain Integration Between Purchasing and Logistics

From the perspective of either the purchasing area or the logistics function, each area's strategy is very closely aligned with the higher order business strategy that they intend to support. However, when you examine the two functional strategies together, the overall alignment between purchasing and logistics is poor. In many cases, it contributes to the destruction of customer value and to an increased total cost of ownership (TCO) for the firm. To achieve integration within these supply chain areas, two broad principles should be followed: *supply chain orientation* and *purchasing to logistics integration*.

Supply Chain Orientation

The broad supply chain organization must appropriately develop, manage, and improve supply chain strategies that drive from a common executive orientation. This executive orientation should focus on a number of areas, but should not have a functional focus. Some critical areas of orientation include relevant value focus, strategic resource allocation, knowledge management, change management, risk management, innovation and transformation, continuous improvement, market orientation, and long-term orientation. This executive orientation provides the platform on which integrated purchasing and logistics processes may be appropriately developed to leverage customer value and TCO.

Purchasing to Logistics Integration

For the supply chain to appropriately support business strategy, a change in how supply chain functions undertake their business processes is required. That is, firms must remove the organizationally convenient but artificially functional silos: purchasing, planning, manufacturing/operations, and logistics. Only then can the underlying integrated business process around supply chain management function properly.

Step-by-Step Process for Integrating Purchasing and Logistics: Eight Steps

The following is a discussion of some of the strategic process areas that may provide the greatest opportunity to improve integration of the critical—but less-integrated—supply chain functions of purchasing and logistics. These strategies are driven by the "executive orientation" described above, and produce the "performance-targeted" outcomes discussed in the following section.

1. *Strategic approach*—The purchasing and logistics functions should co-develop all or parts of their respective functional strategies. In doing so, each function will have greater visibility into the drivers and processes of the other function. This visibility will create an opportunity to remove competing processes and objectives.

2. *Measurement design*—The performance measurement systems within both purchasing and logistics should be appropriately commingled. For instance, to drive toward improved alignment between the functions, purchasing should be co-tasked with logistics performance metrics and logistics should be co-tasked with purchasing performance metrics. This falls under the old adage of "what is measured is what will be valued."

3. *Organizational design*—The organizational structure of purchasing, logistics, and any overarching supply chain function should be examined for opportunities to create improved integration. In many firms, the logistics function and the purchasing function do not report to a common "supply chain" executive. Without an organizational design that supports integrated processes and capitalizes on opportunities from functional integration, it will be difficult to benefit from the alignment of purchasing and logistics.

4. *Partner selection*—Jointly choosing suppliers and logistics service providers may well provide insight that could be leveraged into improved customer value and improved TCO. The choices have implications across the supply chain functions that are important to delivering customer value.

5. *Partner development*—Once a partner selection decision has been made, certain partners will overperform, other partners will meet expectations, and some will underperform. There may be an opportunity to invest in "partner development" with underperforming partners. Given that a firm typically has relatively limited resources for partner development, considering both supply partners and logistics partners simultaneously allows for the most effective use of these limited resources. Furthermore, the development of a logistics partner or supply partner may be done more effectively in some cases with a joint purchasing/logistics approach.

6. *New product development*—Developing products that deliver the greatest value to a firm's immediate customer and the ultimate consumer may be done most effectively when both the supply base and the logistics supplier base are leveraged in areas where partners hold core competencies. In other words, a purchasing/logistics–integrated product/process/service design can be better targeted at the desired value of the ultimate consumer.

7. *Global supply chain management*—The implications for cross-functional integration between purchasing and logistics are

paramount when considering the global nature of consumers, logistics systems, and the location of supplier partners. Designing this global supply chain to deliver the best customer value and lowest TCO can only be accomplished by jointly considering both the logistics system and the characteristics/location of existing and potential suppliers.

8. *Supply chain security*—Firms have increasingly felt the negative effects of risky/unexpected events that occur in the form of natural disasters, supplier failure, terrorism, etc. Designing a robust supply chain should include, among others, the purchasing and logistics functions. However, the purchasing and logistics functions must examine the issue of supply chain security together. Only together can a supply chain design be developed that appropriately understands, prevents, and/or mitigates the risk of supply chain disruption.

The Benefits of Integrating Purchasing and Logistics

Ultimately, purchasing and logistics can work together and with manufacturing and planning to create supply chains that deliver breakthrough performance to both the firm and the firm's stakeholders. Specifically, this performance may come in the form of customer satisfaction, cost/price, TCO, quality, cycle times, delivery, responsiveness, flexibility, innovation, environmental/social sustainability, compliance, and developing/enhancing/protecting intellectual property.

This "performance-targeted" supply chain design is accomplished by leveraging the strategies discussed above, which are driven by an appropriate executive orientation. Certain enablers may need to be present for this "executive orientation" to "supply chain strategy" to "performance-targeted supply chain design" to properly function.

Enablers of Cross-Functional Integration

In order for an executive orientation to drive appropriate integrated purchasing and logistics strategies, certain enablers may well be needed. Some of these enablers may include

- An understanding of the global environment
- An understanding on the part of supply chain functional leadership of both the finance and accounting functions in the language associated with these functions

- An understanding on the part of supply chain functional leadership of the marketing function
- A focus on acquiring, developing, retaining, and offboarding supply chain talent
- An understanding of cross-functional teaming
- An understanding of information systems and technology
- An understanding of supply chain analytics

Summary

This chapter presents our view of the current state of cross-functional integration, and articulates a number of important drivers to make this integration successful. This chapter also develops the importance of the specific case of cross-functional integration between the purchasing and logistics functions. We believe that for too long these supply chain functions have been strategically aligned with their functional strategies and, as a result, have been competing with each other instead of working together. That misalignment leads to the destruction of customer value and reduced firm performance.

While, from a process perspective (plan-source-make-deliver), these functions are separated by the "value-creation" function (planning, manufacturing, service provision, etc.), from an overarching process perspective they are extremely integrated. In fact, they are so highly related that managing them in functional silos is one of the largest factors leading to disintegrated supply chains. Supply chain executives should turn their supply chain integration focus inwards. While the "great divide" was once described as the chasm that existed between marketing/sales and manufacturing/operations, the new "great divide" is between purchasing and logistics. Firms that strive to close this chasm will outperform those that do not. This need for internal integration across the internal supply chain will be covered more thoroughly in Chapter 5.

Absolute Value for the Firm to Relative Value for Customers—World-Class Metrics

By Paul Dittmann, Ph.D., Executive Director,
The Global Supply Chain Institute

Why Is This a Game-Changing Trend?

Can simply changing the performance measurement and goal setting system inside a firm significantly enhance the overall performance of the supply chain? You bet it can. Just as firms need a transformational supply chain strategy, they need a metrics and goal setting system that drives the new behaviors required by the new strategy. The old phrase still applies: "If you always do what you always did, you will always get what you always got." We need a new metrics/goal setting system aligned with a transformational strategy to change the supply chain game.

For example, a large CPG company felt they needed something that would break through the firm's complacency, something that would propel them to a new level of performance. They developed the perfect order metric, calculated by multiplying together four factors for each customer order: on-time, complete, damage free, and invoiced correctly. Instead of basking in the glow of 95% performance in each category, they were stunned to see a new metric that showed an 81% level of performance (e.g., $0.95 \times 0.95 \times 0.95 \times 0.95$). This drove the organization to a totally new level in customer service, caused their competitors to scramble, and significantly increased market share and sales.

Supplier and Customer-Segmented Metrics Can Be Powerful Tools

At the University of Tennessee Supply Chain Forum, one of the most popular topics is the subject of how to measure supply chain performance. Many supply chain professionals seem to be searching for the "magic bullet," or a set of key performance indicators (KPIs) that suddenly create a new, higher level of performance. What will it take to get to the next level in supply chain metrics? Before answering that, let us reflect for a moment on the progress made.

Our 2000 survey indicates that firms are looking at supply chain performance in a more sophisticated manner. For example, the data in the Absolute to Relative Value figure show a strong move toward performance assessment on a customer segmentation basis. The older survey indicted more of a focus on overall versus *segmented* customer measures. Of course, a segmentation approach can benefit the supply side as well as the demand side. Supplier and customer-segmented metrics can be powerful tools.

Even though Figure 1.6 shows encouraging progress, the 6.9 score is the lowest survey score among the 10 megatrends, meaning that much improvement is still possible. In fact, we find that performance

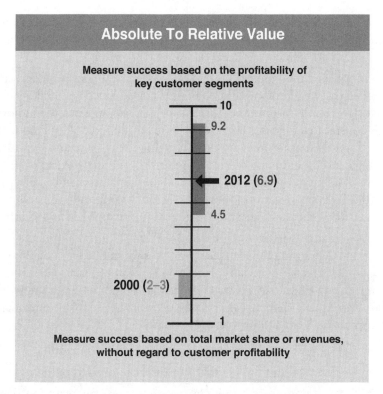

Figure 1.6 Significant progress has been made on metrics—More to do

measurement in many firms is often labor intensive and inconsistent, held together with manual spreadsheets, decoupled from the strategy, and excessively detailed with a false sense of precision. As one executive said, "For us, it seems like death by a thousand metrics." Simple is better for metrics frameworks. Companies need a clear framework that can be translated to all levels of the organization with clarity for all employees.

From our data on hundreds of companies, we believe that best practices for designing a metrics framework do exist. Before discussing those best practices, let us discuss why performance assessment is on the minds of supply chain professionals today.

Mandate for New Metrics

As your supply chain strategy evolves, many of you are asking, "what is new in the metrics world?" In the 1980s, we would have said, "supplier scorecards." In the 1990s, we would have said, "balanced scorecards." Today, what is new is the availability of masses of data from multiple sources, usefully organized in data warehouses to link outcomes with drivers and turn data into true insight.

Supply chain organizations and strategies must continue to evolve to meet the needs of rapidly changing customers, the challenge of aggressive competitors, and the inexorable advance of technology. This constant change demands a continual evaluation of existing metrics, if not a completely new set. Changing the organization and the strategy and then relying on the same tired set of metrics makes little sense. The right supply chain KPIs aligned with the right accountabilities help the organization deal effectively with tradeoffs and proactively drive the kind of behaviors that support supply chain strategy.

In a recent survey of our Supply Chain Forum members, we found that "choosing the right metrics" was rated third among 25 possible topics that supply chain professionals want to learn more about. Many supply chain executives thirst for a better metrics framework. They worry that their existing KPIs prevent them from optimizing performance. They tell us that they want to:

- Learn how excellence is achieved in similar companies
- Understand the drivers that most impact needed outcomes
- Set appropriate performance measures and targets for improvement
- Learn how others enable and empower employees to make change happen
- Understand how to create a culture of continuous improvement

In another of our surveys, 34 business executives from a broad range of companies ranked the following supply chain issues on a scale of 1 to 10, with 10 the most important and 1 the least important. With an 8.15 rating, performance measures and goal setting as the most important issue should not have surprised us, but it did (see Table 1.2).

Based on our database of hundreds of companies, we believe that best-in-class firms establish a metrics framework using four key principles, which are discussed in the remainder of this section:

1. Create the right cross-functional accountability
2. Establish a driver-based metrics framework
3. Set appropriate goals
4. Ensure that metrics cannot be easily gamed

Table 1.2 Executives Feel Strongly that SC Measures and Goal Setting is Our Most Important Work

Ranking of Supply Chain Executives' Interests	
Implementing the right metrics and setting the right goals	8.15
Establishing collaborative relationships with suppliers and customers	7.91
Advances in supply chain visibility	7.80
Professional development, training, education	6.71
Helping with revenue generation	6.62
Managing the global supply chain	6.55
Effectively using technology	5.21

Create the Right Cross-Functional Accountability

Measuring something accomplishes little unless the right accountability is established. Good supply chain leaders should always ask themselves if their metrics have been designed with the right cross-functional accountability in place. For example, the accountability for inventory, forecast accuracy, and product availability should be shared between the supply and the demand sides of the organization. One executive told us that only the production planning function had the goals for inventory turnover in their personal performance plan. Yet the planning function controlled neither the input to inventory (manufacturing) nor the output (sales). In this case, planning had all of the accountability and none of the control.

Unfortunately, this situation is all too common. There are few companies in which sales shares accountability for inventory. Yet sales strategies tremendously influence inventory levels. This particular issue is one of the greatest organizational accountability flaws in firms today.

Establish a Driver-Based Metrics Framework

Are your metrics linked in a logical framework to your overarching goals, or are they simply a laundry list of items with no apparent logic? If the prime goal of the firm is to drive shareholder value, then a framework needs to be established so that the individuals in the organization can clearly see how every sub-metric flows into shareholder value.

We believe that the best practice is to first list your key outcomes. Then you will be able to identify the drivers of those outcomes.

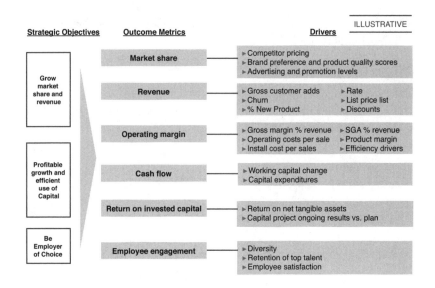

Figure 1.7 A strong tie between strategy and measures is required. This enables the strategy to live

According to research sponsor Ernst & Young, leading companies use statistical correlational analysis to find the drivers that best link with the required outcomes. Ideally, they find drivers that have a disproportionately positive impact on the big outcomes needed. They then set up a hierarchical framework, or "driver tree," to visually show how each metric feeds overall goals, as shown in Figure 1.7.

The expanding availability of data organized and linked in a data warehouse makes it possible to find these correlations. In the future, data will be linked across the entire supply chain, from suppliers to customers. In the old days, companies were limited to data that existed within their four walls. Now it can be integrated across multiple parties in the value chain to measure the performance of extended supply chain.

In another example, a manufacturing firm we assessed selected "the efficient perfect order" as its ultimate goal. The basic perfect order performance is calculated by multiplying together performance in four areas: on-time, complete, damage free, and invoiced correctly.

The "efficient" perfect order in this firm then included cost and inventory as key additional factors. With the efficient perfect order as the overarching goal, this firm built a metrics driver-based framework, illustrated in Figure 1.8, to show how all of the sub-metrics contributed to achieving the perfect order.

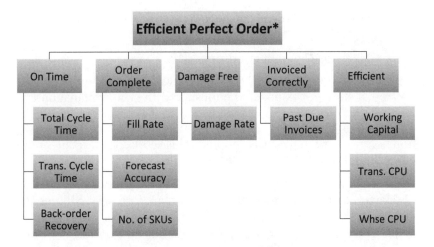

*Figure developed at the University of Tennessee for classroom instruction. Later referenced in the book *Supply Chain Transformation*, 2012, McGraw Hill, J. Paul Dittmann, page 187

Figure 1.8 Focus on perfect orders requires broad SC capability improvement

The Perfect Order Metrics Framework

A clear driver-based metric framework creates a common business language that leadership can leverage to align the organization toward common goals. One supply chain senior executive at a leading retailer consistently stresses the need for three outcomes in exactly the same order: product availability, inventory, and cost. He wants all metrics focused on driving to world-class levels in each of these big three outcomes.

In developing the new metrics to support your strategy, make sure you have a driver-based framework with a set of criteria in place to avoid poorly designed or seriously flawed metrics. For example, we worked with one firm that defined a set of excellent criteria to design the new supply chain metrics needed to support its supply chain strategy. These characteristics became a hurdle test. Metrics had to reasonably satisfy the following criteria to be part of the KPI framework:

- Stable and accurate data with few, large, random, or unexplainable swings
- Understandable to everyone, along with a "line of sight" so that key personnel can see how their actions influence the metric
- Designed so that they cannot be easily manipulated or gamed

- Capable of drill-down analysis so that the root causes of changes are apparent
- Clear cause and effect drivers
- Easily accessible for relevant parties and available in clear reports that were developed and published with clear explanations

These criteria always generated good discussion in the organization before a new metric was adopted. This resulted in a smaller number of high impact KPIs.

Finally, a firm's supply chain metrics need to be supported by a disciplined and documented metrics governance process. Metrics governance is not a very exciting topic, but it can be the most common and largest obstacle to execution. Companies need to define consistent taxonomy, data standards, driver libraries, and metrics definitions.

Set Appropriate Goals

Selecting the right metrics and defining the associated responsibilities is very important. Establishing goals is an entirely different matter. Too many companies only use internal comparisons (year over year performance, for instance) and feel good about achieving an internal goal. This "comparison of yourself to yourself" is a very dangerous practice. For example, one consumer product manufacturer achieved a 6.7 inventory turnover level on its finished goods inventory, a 15% improvement above the 5.8 level from the prior year. Unfortunately, when doing a competitive assessment, the company discovered that its major competitor had achieved an 8.5 inventory turnover level. The 15% improvement did not look so good in light of that statistic.

Ensure That Metrics Cannot Be Easily Gamed

Many companies "game" their metrics. Supply chain professionals rationalize this data manipulation (e.g., "It would be unfair to include that SKU in our fill rate calculation; we have had supplier problems and cannot get enough of that product"), but in the end it only hurts the supply chain organization. This is because it hides real performance and creates a disconnect between the company's perception and the experience of the customer.

In a recent survey we conducted, 81% of respondents believe their company provides superior customer service. Yet only 8% of customers say they receive superior customer service. Likewise, in a recent analysis of our database, 94% of firms rated themselves above average in satisfying their customers. Since it is statistically unlikely for 94% of companies to be above average, these respondents are either manipulating or overestimating their capability. Overestimation is more than just naïve; it actually destroys internal motivation because employees hear how well the firm is doing and feel no urgency to surpass competitors or delight customers, thereby giving rivals an upper hand.

Example:

- A supply chain vice president (VP) in a CPG company described the extreme pressure he faced at all levels of his organization to deliver better fill rates. He said that the sales organizations continually communicated horror stories where customers bitterly complained about not being able to get product. "The CEO called me one day and made it quite clear that fill rates had to improve. In fact, he demanded that large three by four feet charts be posted in prominent places around the building to show the improvement in fill rates that must come about!" He then related how, during the subsequent weeks, the company struggled with manufacturing and vendor issues, which offset any internal fill rate improvements. The pressure on the supply chain organization built to an excruciating level.

    Suddenly, everything changed. The metrics started to show fill rate improvements, which continued until the company's goal was achieved. The supply chain VP was amazed. He was also confused because customer complaints continued unchanged. Much later, his director of inventory management left for another company and the replacement discovered that his predecessor had directed the inventory analysts to exclude data when they calculated fill rates.

    When new products entered the system, it took several months for the inventory to catch up with demand. This imbalance negatively impacted the fill rates at the company, so the inventory analysts decided to eliminate that data from the fill rate calculation until the new product stabilized. They did not tell anyone, rationalizing that they were making the metric more accurate by eliminating such detail. This slippery slope became steeper and steeper, and the analysts began manipulating other "unfair situations." Eventually, the house of cards came crashing down. Several of these analysts were dismissed from the company, and the supply chain VP had to explain the abrupt and embarrassing fall in the fill metric once the data were corrected.

Summary

In summary, we believe you should design a new set of supply chain metrics that support the supply chain strategy, follow a logical framework, clearly define cross-functional accountability, relate to set goals with best-practice benchmarking, focus on customers, resist gaming, and provide effective insights into how the supply chain organization is performing and where improvements can be made. In summary:

1. Create the right cross-functional accountability.
2. Establish a driver-based metrics framework.
3. Set appropriate goals.
4. Ensure that metrics cannot be easily gamed.

If your organization manages to do these things, it is taking the necessary steps to make the jump to world-class metrics.

Forecasting to Endcasting—Demand Management

By Mark Moon, Ph.D., Department Head, University of Tennessee, Department of Marketing and Supply Chain Management

Why Is This a Game-Changing Trend?

Few supply chain professionals are pleased with the forecasting process in their firms. We often hear the familiar refrain, "If we could only improve forecast accuracy, most of our problems would disappear." That is very likely not true. Instead, the real issue is how the forecast plays in the cross-functional demand/supply integration process and how the firm truly manages demand. We rarely see firms with a well-defined, rigorous process for managing demand. The few with those processes make game-changing transformations in their business.

John Deere is a great example of a company that has changed the game with a world-class forecasting and demand management process. In presentations at our Supply Chain Forum, Deere described how improvements in this process led to major simultaneous improvements in cost, working capital, and product availability. While many factors have led to John Deere becoming a great company, world-class forecasting and demand management process is definitely a critical factor.

Improvement, But Still Short of Maturity

No one buys a company's stock because of its forecasting skills. A company needs to do something with that forecast. They need to become adept at demand management and then translate that into higher revenue, lower cost, and higher cash flow. What will it take to get to the next level in managing demand? Before answering that, let us reflect for a moment on the progress made (see Figure 1.9).

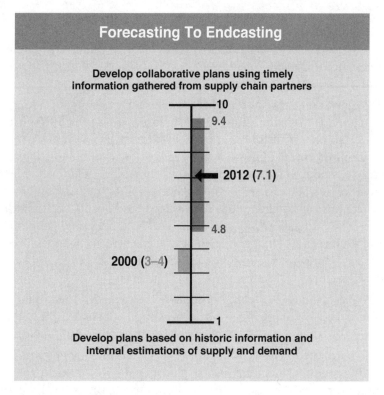

Figure 1.9 Significant work on forecast improvement but we are not to endcasting yet

In the 2000 landmark study, the authors predicted that the trend from forecasting to "endcasting," (or "demand management") was a relative laggard of the 10 game-changing trends. This trend achieved a 4.5 maturity score on a scale of 1 to 10, with 1 representing no meaningful acceptance and 10 being total adoption. The prediction was that, by 2010, organizations would be operating close to the total adoption level. The research supporting this chapter suggests that the industry has improved substantially on this trend but has much opportunity for improvement.

The firms' average response to the 2012 survey was 7.1, with a range (at one standard deviation) of 4.8 to 9.4. The important takeaway from the data is that considerable progress has been made, with considerable movement toward firms developing collaborative plans with timely information gathered from supply chain partners. However, with a mean score of 7.1, much opportunity still exists for greater maturity in collaborative forecasting with supply chain partners.

The State of Forecasting Today

Forecasting as a process has evolved rapidly, given swift innovations in data processing for easy storage, retrieval, and high-speed processing of detailed or granular level business data through complex forecasting algorithms.

According to research partner Ernst & Young, organizations are no longer content with using univariate forecasting, which simply uses sales or shipment history as a basis for generating the forecast; they are adopting the use of "scenario-based forecast simulations" using multiple causal variables. These input variables include macroeconomic factors such as segmented demographics, population growth, buying patterns by age group, promotions, and events such as 4th of July or the Super Bowl. All these are used to process and generate an accurate forecast in the near term (4 weeks) and a sustainable forecast in the medium (3–6 months) and longer term (2–5 years) to drive their downstream supply chains.

What is game-changing in the forecasting and demand planning realm is the ability of high-end forecasting processes to "sense and respond" to dynamic and evolving customer or consumer behaviors.

According to Ernst & Young, to do this it requires mining data from multiple sources, including social media, public blogs, and relevant subscription-based services (IRi, Neilsen, Dun & Bradstreet), weather data, labor statistics, and then modeling the data through analytical engines. The data are then processed and made relevant to the forecasting process to predict the sales of goods to customers at the right time and location (demand point). This demand signal is then rippled through the planning and decision support processes such as Integrated Sales and Operations Planning and then through the actual execution from sales order processing to manufacturing execution.

Offshoring and Long Lead Times Increase Priority

While the original research pointed to close collaborative relationships with supply chain partners to enhance forecasting effectiveness, one macrotrend has made prediction of future demand, whether that is thought of as "forecasting" or "endcasting," considerably more difficult—and considerably more important. That trend is the relentless move to offshore manufacturing to low-wage countries, particularly in Asia, but also in Latin America. One consequence of this offshoring is a lengthening of lead times for many manufactured goods.

When forecasting and managing future demand, *the horizon (or the length of time into the future that demand is being forecasted) must be at least as long as the lead time.* When manufacturing is moved offshore, forecasters need to be thinking 4 to 6 months into the future and beyond instead of the traditional 2 to 3 months. *The longer the forecasting horizon, the less accurate the forecast will usually be.* A "guess" about what is likely to happen next month will be better than a guess about what is likely to happen 6 months from now.

Thus, one of the consequences of the trend to offshore manufacturing to low-wage countries is that the forecasts that are necessary to drive these longer lead times are usually less accurate than they would be if the manufacturing were taking place closer to the customer. When accuracy is lower, more inventory is needed to deliver acceptable service levels to customers. It is in this way that offshoring has made forecasting more difficult. Consequentially, it has also made forecasting more important. Long lead times result in more inventory in process throughout the supply chain, and if the forecasted demand is either way too high or way too low, the risk of obsolescence, or unhappy customers, is greater.

Internal Collaboration and S&OP

While many companies have improved in their ability to collaborate effectively across organizational boundaries with supply chain partners to improve forecasting effectiveness, a greater effort has been made to enhance internal collaboration among the demand, supply, and finance functions to transform accurate demand forecasts into actionable business and supply chain plans.

No matter if the process is called Sales and Operations Planning (S&OP), Integrated Business Planning (IBP), or Demand/Supply Integration (DSI), it is designed to match future demand projections (demand forecasts) with future supply capabilities (supply forecasts). Doing so allows the organization to formulate business plans that achieve a balanced set of goals. While any company would agree that these are worthy goals, transforming siloed organizational cultures into cross-functional, collaborative ones has proven to be elusive.

A variety of culprits are to blame for the lack of success in these processes, but failure to engage the demand side of the enterprise (sales and marketing in a manufacturing context, and merchandising in a retailing context) is by far the greatest barrier to successful S&OP implementations. Accurate forecasts are not worth the paper they are written on if they are not a part of a robust Demand/Supply Integration process. Transformation of demand and supply signals from external supply chain partners into collaborative business plans is the desired end state. Such transformation requires a collaborative organizational culture, a robust set of processes, and effective information technology tools. Considerable work still needs to be done in many companies to achieve this ideal.

Forecasting Versus Endcasting/Demand Management

In the 2000 landmark study, "forecasting" was conceptualized as predicting the demand from a supply chain partner. "Endcasting" was seen as a heightened focus on final consumer demand. CPFR was discussed as a movement toward endcasting, in the sense that it represented a collaborative effort between retailers and manufacturers to model and then manage consumer demand.

A macrotrend that has helped further this ideal of "endcasting" is the availability of vast amounts of end-user demand data. With the ubiquity of scanners, data are now available for analysis and insight into demand patterns not just from channel partners but also from end consumers.

Terra Technology, a leading vendor of demand management software, is seeing companies actively take advantage of big data through automated analytical tools with their Demand Sensing product. In their Forecasting Benchmark Study (encompassing one-third of North American CPG traffic from multinationals such as Procter & Gamble, Unilever, Kraft, Kimberly-Clark, and General Mills), companies that use big data to sense demand reduced forecast error by 40%. As part of their demand-driven journey, General Mills cut 7 days of inventory between 2011 and 2012 (source annual reports). These are game-changing jumps in performance that can turn into considerable financial gain.

For example, business forecasters have traditionally gravitated toward time-series methods to identify historical demand patterns that repeat with time. One of the most widely used time-series tools is the Holt–Winters approach to exponential smoothing, which was first developed over 50 years ago. Few useful time-series approaches have been developed since Holt–Winters, and with the exception of demand-sensing applications, forecasters have generally lagged behind their business analytics colleagues in the use of sophisticated data tools to identify causal patterns of demand. Such causal modeling tools are more important today than ever, with the emergence of new forms of promotional tools and the growth of "shopper marketing" as a critical marketing strategy.

While time-series tools are far less useful in a promotion-intensive environment, demand-sensing technology has been shown to be effective. During promotions, the same benchmark study referenced above found that the companies that sense demand reduce error by 34%, allowing them to sense consumer response to promotions and better serve customers during these high-profile events. Similar performance is experienced for new items, with Demand Sensing reducing error by 32%.

Keep in mind, though, that regardless of what it is called, it is a guess about future demand, and it will always, always, *always* be, to some degree, wrong! Best practice is to gather as much information as possible from historical demand patterns (statistical forecasting), current manufacturer and retailer data from nodes across the supply chain (demand-sensing pattern analysis), and individuals who have insight into how the future might differ from the past (human insight). Thus, there are multiple sources of input for every forecast. How do you make sense of it all? Best practice is to relentlessly measure the accuracy of the inputs provided from different sources and to use those measures of accuracy to judge the added value from each input source. As an example, software exists that measures the predictive value of each input source at every item's location and uses the most predictive signals to create the best possible forecast. It gets complex very fast, so it is a good task for software.

Example:

- Company D, a global manufacturer of agriculture and construction equipment, collects forecast information from a variety of sources. It conducts extensive statistical analysis on historical demand patterns and then, using sophisticated tools, projects those patterns into the future to create a baseline forecast. Company D then asks its sales teams to adjust these baseline forecasts when they have insight from their large customers that previous demand patterns might change. It asks its marketing teams to adjust these baseline forecasts when they have insights about new product introductions, new market entries, or pricing actions. The company also collects detailed forecasts directly from its large customers.

    The forecasting team at this company compiles each of these inputs on future demand, and measures the contribution made by each. Did the sales teams' adjustment make the forecast better or worse? Did the marketing team make the forecast better or worse? Did the customer forecast add value? Based on these accuracy metrics, each input is assigned a "weight" at the consensus forecast meeting, and the consensus forecast is created by applying this weight to each input. If the sales teams want their forecasts to be taken seriously, they have to demonstrate that their adjustments add value. This is an example of how multiple perspectives can be systematically and analytically used to create accurate and unbiased forecasts, which becomes a critical component of a global S&OP process that has resulted in lower inventories, lower operating costs, and higher customer fill rates.

Summary

There have been exciting improvements made in forecasting and demand planning processes in the 15 years that have transpired since the start of this research stream. Firms have begun to realize that the statistics applied to the process may not be as important as the process itself, and have adjusted accordingly. As with so many of the game-changing trends, however, much work remains to truly achieve a desired state of demand planning that optimizes supply chain performance.

Training to Knowledge-Based Learning— Talent Management

By Paul Dittmann, Ph.D., Executive Director,
The Global Supply Chain Institute

Why Is This a Game-Changing Trend?

Few would argue that the number one requirement for supply chain excellence is finding, hiring, and developing talented people and then placing them in the proper roles. Firms that are good at this clearly have a competitive advantage. To change the supply chain game, you have to be world class at getting the right people with the right skill sets in the right places, and that does not happen without a lot of effort and focus.

During the Great Recession, talent management took a backseat in many firms. However, one large manufacturer of consumer and industrial products decided to move aggressively against this trend. They maintained the training budget when all else was being cut. Even when salaries were reduced 10% across the board, they maintained their commitment to developing their people. They also resolved to transform their workforce, even though they had to cut headcount. They aggressively sought and hired the best and brightest talent they could find. This company believes that, when its competitors were retrenching, it built a lasting foundation of talented people for decades to come. We tend to agree.

Education Is Being Focused on Broad Goals of the Firm

If supply chain professionals had to choose one element to master to improve their performance, they invariably tell us it would be talent management. Our industry partners say that finding, hiring, and developing supply chain talent is the number one requirement for transforming a supply chain. What will it take to get to the next level in managing supply chain talent? Before answering that, let us reflect for a moment on the progress made.

In our survey of over 160 supply chain professionals, we found a significant improvement in the way firms are developing supply chain talent. From Figure 1.10, we can see that companies have clearly moved from an exclusive focus on training only for specific functional skills and have added more education focused on helping their people improve overall organizational performance.

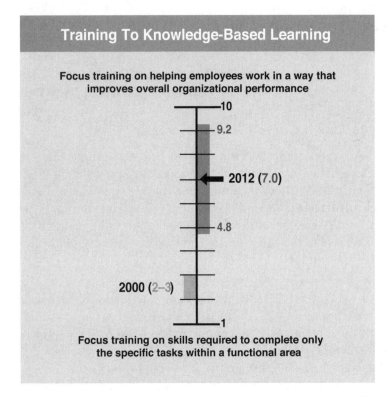

Figure 1.10 Progress has been good on our transition to learning versus teaching

However, training and education to develop talent is putting the "cart before the horse." Companies first need to acquire the right talent, and that means they have to define the skills and competencies needed by the modern supply chain executive. More firms have an increased expectation from supply chain leadership. For now, suffice it to say that the supply chain, more often than not, is the prime driver of shareholder value (or owners' equity in firms). Yet most companies fail to fully leverage this pathway to driving value, often because they lack the right talent to make it happen.

The Modern Supply Chain Executive

Fifteen years ago, the supply chain leaders in most companies held a title such as "vice president of logistics." This largely functional role relied on technical proficiency in discrete areas: knowledge of shipping routes, familiarity with warehousing equipment and distribution-center locations and footprints, and a solid grasp of freight rates and fuel costs. The VP of logistics reported to the chief operating officer or chief financial officer, had few prospects of advancing further, and had no exposure to the executive committee. The way companies need to think of the modern supply chain executive, however, has changed dramatically.

Supply chain executives still need to be experts at managing supply chain functions such as transportation, warehousing, inventory management, and production planning. But the supply chain process extends end-to-end and even outside the firm, including the relationships with suppliers and customers on a global basis. Leading firms now see the supply chain functional leaders as the executive to coordinate the end-to-end supply chain process, even though they do not control it all. Because of that added dimension of cross-function, cross-company coordination, senior supply chain executives must possess a number of unique characteristics.

Supply Chain as Part of the Executive Team

Today, in a growing number of firms, the supply chain chiefs of high-performance companies do not just have access to the executive team—they are part of it. That role requires not only educating the CEO and the Board by giving them the vocabulary to talk about supply chain subjects and its critical role in creating economic profit, but also finding and driving opportunities to increase economic profit. The job in those progressive firms is no longer a mostly functional one, but instead plays a key strategic role that can influence 60% to 70% of a company's total costs, all of its inventory, and most aspects of customer service.

The supply chain leaders in these progressive firms have global responsibility for coordinating processes across functional silos such as sales, research and development (R&D), and finance, as well as functional responsibility for activities such as procurement, logistics operations, production planning, and customer service. They pay as much attention to the demand side as to production and materials planning, and know what it takes to reliably deliver products to customers and to build mechanisms to learn what customers have to say.

Finding talented people who can grow into these roles is, according to one of our surveys, the most challenging aspect of supply chain people management. Developing, retaining, and evaluating talent are also daunting tasks, but simply finding and successfully hiring top talent seems to be the most challenging of all.

Consistent with our research, the "Chief Supply Chain Officer Report," published by *SCM World* in October 2012,[5] reported that the percentage of respondents who thought certain aspects of talent management "extremely or somewhat challenging" were:

- Finding talent: 83%
- Hiring talent: 77%
- Developing talent: 66%
- Retaining talent: 66%
- Measuring talent: 47%

It should be no surprise that finding and hiring top talent is at the top of the list. Demand for the most talented supply chain professionals will continue to rise, and hiring and retaining them will continue to tax the best organizations. Companies must "sell the opportunity" to candidates much more adroitly. So what must firms do to ensure that they have the supply chain experts who can drive true strategic success? First, they must create the right environment and then acquire the very best supply chain talent available. Finally, they must aggressively develop that talent. Finding top talent starts with identifying its critical characteristics. This process is addressed in depth in Chapter 6.

The Critical Competencies of Top Supply Chain Talent

The best talent can only be acquired after it has been identified. To select the right people to oversee the increasingly pivotal supply chain responsibility, firms must know the blueprint for the "dream" supply chain leader. Our recent book, *The New Supply Chain Agenda*, groups these characteristics into five key competencies[6]:

- Global orientation
- Cross-functional, cross-company understanding
- Leadership skills
- Technical and analytics savvy
- Superior business skills

---

[5]"Chief Supply Chain Officer Report," *SCM World* (2012, October).
[6]Slone, Mentzer, and Dittmann, *New Supply Chain Agenda*.

Embedded in all of these is the need for a strong focus on relationship management skills with internal customers, employees, and external partners.

Global Orientation

Nearly all senior business executives today need to be globally capable. Global sourcing and global supply chains have expanded tremendously in recent years, for both retailers and manufacturers. There are few companies or competitors that do not either source or sell globally. Therefore, supply chain executives must manage an enterprise that extends across continents and must deal effectively with suppliers and customers worldwide.

Global executives we talk with say that resources to help them learn all of this are few and far between. Just being aware of the myriad rules and regulations is a daunting task. Most have learned via experience. And, in fact, this may be the only effective way to truly internalize global knowledge. That is why companies like P&G require their high-potential employees to have a global assignment before they can be promoted to the executive level. While this may not be possible for all companies, it is the reason why supply chain executives with global knowledge are so extremely valuable.

Cross-Functional, Cross-Company Understanding

Unlike some other senior executives, supply chain executives must embrace the added dimension of cross-functional and cross-company complexity—the challenge that comes with thinking of the end-to-end supply chain as an integrated system. Manufacturing or sales executives must develop deep expertise and a strategy for their area. But supply chain executives must also comprehend the connections and interdependencies across procurement, logistics, manufacturing, and marketing/sales. In addition, they must also absorb the complexity of interfaces outside the firm, with suppliers and customers.

Leadership Skills

A growing number of today's supply chain leaders are front and center within the organization. They must be able to foster close interpersonal relationships that build credibility for them and for the function across the organization. They must be able to build teams and manage people and must communicate their message compellingly to multiple stakeholders. They find themselves in the position of having to influence others in the firm to work together to create a world-class supply chain. To get the whole picture, they are masters at building close collaborative relationships with their companies' leaders in sales, marketing, human resources, and finance.

Technically and Analytically Savvy

Technology has become a key enabler of supply chain excellence, and spending on supply management applications and services continues apace. Indeed, in our work across hundreds of companies (including retailers, manufacturers, and service providers), we almost always see that supply chain often consumes the majority of the IT spend in firms, with that spending supporting warehouse-management systems, transportation-management systems, inventory management, and production planning systems. The supply chain chief need not be credentialed in IT systems or other technology areas, but he or she must know what to avoid and what questions to ask to successfully guide the implementation of new supply chain technologies.

The need for analytics capability is exploding. The expanding role of the supply chain in most firms leads to an exponentially higher level of complexity and therefore an expanded need for analytics and modeling skills. Yet such skills are in alarmingly short supply. According to a McKinsey & Company study, by 2018, the United States could face a shortage of 1.5 million managers and analysts with the know-how to use big data to make effective decisions. Academic institutions and companies must partner to find ways of developing these skills to meet the ever-expanding demand for them.

Superior Business Skills

Supply chain leaders must be business people first and supply chain specialists second. Their foremost focus must be on enhancing economic profit and shareholder value, not simply on cost cutting. Supply chain leaders must be able to speak the language of senior executives as easily as they can talk about fleet-truck efficiencies or demand forecasting. Terms such as EBITDA (earnings before interest, taxes, depreciation, and amortization), ROIC (return on invested capital), and shareholder value should be part of their everyday parlance, and supply chain leaders should be as comfortable discussing cash flow with the treasurer's office as they are with talking about delivery schedules with suppliers. Supply chain issues are often the least understood by the Board and the CEO, and they must be explained in executive language.

Acquire the Best Talent

After carefully determining the right mix of functional proficiencies and the right combination of the five universal characteristics discussed above, companies must enter the contest for talent. Engaging in the battle for scarce talent involves viewing the world, other industries, and supply chain management training programs as the talent basket. Supply chain management is no longer limited by national borders or industry boundaries. Leading practitioners consider the world their "talent basket." Their searches extend accordingly to India, China, Brazil, Europe, and beyond.

From Training to Knowledge-Based Learning

Finally comes the need to develop talent for key supply chain process roles. This involves creating a professional development plan for every manager in the supply chain organization. Too many supply chain managers lack sufficient knowledge of how the rest of the company runs. Leading firms think creatively as they create ways to develop the talent needed. Any global approach to talent will not succeed easily; it will require a major effort, involving detailed planning of an array of leadership development initiatives.

Universities are especially stepping up, not only with more appropriate education for the growing numbers of aspiring supply chain leaders but also with executive education programs that help shore up the business savvy of established supply chain specialists. When it comes to functional expertise, companies can do more to align and drive new phases of supply chain education at all levels. And there is more room for universities to go further in providing more universal supply chain management skill sets.

Our survey results (detailed in the chart at the beginning of this section) show a huge improvement in the focus on education to improve overall performance of the firm instead of only narrow training in a specific area of expertise. Companies, of course, still need to offer these specific training programs, but it is encouraging to see that firms are moving beyond that to knowledge-based learning. There is still a long way to go, but companies have taken a major step forward in the past 12 years.

We hear from more and more companies that they are sending employees to training but are unsure how to reinforce and internalize the education so it sticks and is used. Leading companies are focusing on post-training programs to make sure their people apply the skills learned. Some of these firms line up projects right after the education session so that the skills learned can be employed. The idea is to move from simply "checking the box" on a training objective to truly seeing a business benefit.

Summary

Few would argue that acquiring, developing, and retaining the right talent is a critical element in building a world-class supply chain. Finding supply chain talent is a special challenge due to the cross-company, cross-functional challenges that need to be embraced. Therefore, the five key talent characteristics discussed above are even more critical for supply chain executives. A talent plan is clearly an essential part of the strategy to drive supply chain excellence.

Vertical Integration to Virtual Integration

By Wendy Tate, Ph.D., at University of Tennessee

Why Is This a Game-Changing Trend?

One of the fundamentals of business is to stick to what you do well and leave the rest to world-class service providers. But outsourcing to a third-party service provider should be based on a distinct strategy to optimize the leverage they provide. Third-party experts can help you change your supply chain game only if you follow a well-defined set of best practices.

For example, a manufacturer and distributor of food products decided to outsource distribution-center operation to a third-party supplier. After the first year, neither party felt good about the performance. There had been no improvement and no savings. But then the firm decided to employ a vested outsourcing model.[7] With this new model, the firm restructured the 3PL contract to pay for results, not just activities, and they provided further win-win incentives for the 3PL to help them make improvements. This allowed the firm to reach new

---

[7]K. Vitasek, M. Ledyard, and K. Manrodt, *Vested Outsourcing: Five Rules That Will Transform Outsourcing* (New York, NY: Palgrave MacMillan, 2010).

heights in serving their customers with less cost and inventory. On-time delivery improved from 80% to 98%, lead times were cut by 55%, and together the two parties shared the $22 million of cost taken out of the operation.

Leveraging the Expertise of Third-Party Suppliers

Figure 1.11 shows significant progress in leveraging the expertise of third-party suppliers as the outsourcing trend advances. What will it take to get to the next level in using third-party providers for maximum advantage? Before answering that, let us reflect for a moment on the progress made.

This trend was definitely in play 13 years ago. For more than a decade, firms have continued to move to more of a virtual integration model with more third-party materials and service suppliers. Virtual integration entails developing strong partnering relationships with material and service providers and leveraging mutual capabilities and

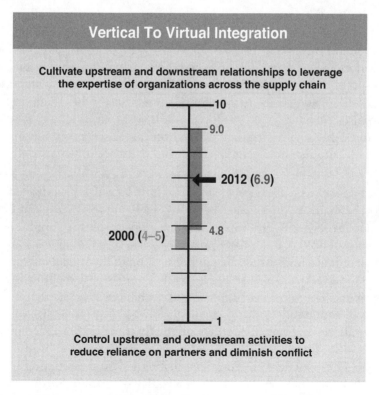

Figure 1.11 Leveraging third party expertise improves your SC's scale

benefits through long-term contracts, information sharing, gain and risk sharing, etc. Meanwhile, they have moved gradually away from the old vertical integration models. But this has not happened across the board. Many organizations have remained vertically integrated to reduce reliance on others and to reduce conflict. Firms have stayed vertically integrated because it provides a number of benefits, including retention of intellectual property, better control of costs, quality, and delivery. Vertical integration also allows an organization to better retain institutional knowledge and focus on competencies.

But the outsourcing trend continues to advance due to the many vertical integration problems—which include the considerable capital investment required, a highly complex organizational structure, and the lack of an available and dedicated workforce. Vertical integration may also inhibit innovation and other market opportunities for an organization.

The Benefits of Virtual Integration

Reliance on an outsourced third-party relationship with material and services suppliers helps to overcome the financial burden of capital investment. It also opens access to a larger pool of skilled labor and management. The benefits of virtual integration include the potential for reduced prices, decreased labor costs, increased knowledge, and access to a larger customer base. Virtual integration can also introduce risk, supply chain disruption, fluctuating costs, and variable quality. Virtual integration may also require an investment in supply chain relationships that causes a shift in power and a loss of control.

In 2000, vertical to virtual integration scored an average of 4.5 level of maturity on a scale of 1 to 10 (10 being total virtual integration to leverage the expertise of organizations across the supply chain, 1 representing total vertical integration). The prediction was that, by 2010, organizations would be operating close to the virtual integration level, or a fully outsourced model, with a pervasive use of third-party suppliers. The research supporting this chapter suggests that industry has continued to move gradually toward virtual versus vertical integration—but not at the rate originally thought. What happened to slow the trend? Will this trend actually reverse itself? There have been highly publicized issues introducing risk and disruption to the supply chain that have caused organizations to continuously rethink and revise the virtual versus vertical integration strategy.

Supply Chain Disruptions are Common

We do not want to confuse a discussion of outsourcing in general with global outsourcing in particular. But global outsourcing especially can demonstrate the risks involved in some outsourcing relationships. None are more sobering than those caused by natural disasters. Natural disasters like the 2011 earthquakes in Japan or the floods in Thailand are somewhat predictable in the sense that Japan is a volcanic region and Thailand is prone to flooding. But the *scale* of the 2011 events proved more extreme than most risk managers could account for. The Honda factory in central Thailand was under 15 feet of water at the high point of the flooding. This incident was minor compared to the March 11 earthquake, tsunami, and subsequent nuclear crisis that engulfed Japan. Toyota suspended production of the Prius in Japan after this event, losing 140,000 badly needed vehicles.

Hundreds of other companies faced major disruptions to their supply chains from this disaster, some lasting through the end of the year. Boeing experienced major delays as a result of the tsunami because the impacted Japanese suppliers produce 35% of the Boeing 787 components and 20% of the Boeing 777 components. General Motors had to halt production in several plants due to shortages from Japanese suppliers. Honda faced severe problems because 113 of its suppliers were located in the affected region of Japan. These twin disasters in Asia in 2011 produced an estimated $240 billion in losses.

Given events such as this, many firms have concluded that a balance of the two extremes is the most logical solution. Rather than outsource everything, they outsource and offshore services and materials in the geographic places that make the most sense from both a total cost, total risk and a customer value perspective. They now understand that effective virtual integration requires the cultivation of upstream and downstream relationships across different geographies to mitigate the risk of increasing labor and investment costs and continuing problems with natural and unexpected disasters.

Virtual Integration of Goods: Outsourcing

For more than a decade, there have been many reports of manufacturing being outsourced, offshored, near-shored, and, most recently, reshored. But for the reasons enumerated above, companies are taking a different view and trying to understand the "right shore," or the right place for their supplies of manufactured goods. Yet the pull of extremely low labor costs and relatively low transportation costs continues to drive many companies to consider manufacturing and outsourcing to suppliers located in other regions of the world, including Africa, South America, the Middle East, and India.

Virtual Integration of Services

At the turn of the century, there was a significant trend in services or knowledge process outsourcing and offshoring. Many staff and design activities were being outsourced to consultants. Information design, collection, maintenance, and analysis were outsourced to experts in information integration. Customer contact centers, R&D, human resource, and even purchasing functions were outsourced to areas with a pool of skilled laborers who were willing to contribute to necessary cost-reduction efforts. Technology integration played a key role in virtually integrating these services suppliers. These trends for outsourced services continue, but they have slowed somewhat for many of the reasons stated above.

Current Trends and Implications

Will the outsourcing trend actually reverse in the future? There is evidence that suggests that forces may be shifting, and in fact some companies are bringing both manufacturing and service production back home. The question that executives are asking today is, "What makes sense for our organization? What are the key considerations in making a location decision? Will there be any new preferred offshore locations, or are some manufacturing jobs and suppliers starting to move closer to the geographic regions where customers/consumers are located?"

The 2008 to 2010 Great Recession motivated companies to reevaluate their global supply chain strategies. Contributing to this assessment was the rapidly rising cost of labor in emerging economies, high oil prices, increasing transportation costs, and a growing awareness of global risks. Beyond higher total cost and increased risk, many firms cite intellectual property erosion and product-quality problems as the underlying reason that some offshore regions are falling out of favor as the low-cost manufacturing locations of choice. Other factors influencing the decision to move closer to home include increased supervision and training on manufacturing and inspections, higher local security needs, and extra expenses associated with travel and telecommunications. According to recent Council of Supply Chain Management Professionals (CSCMP) sponsored research, 56% of offshoring companies are experiencing unexpected increases in total landed costs.[8]

The CSCMP also found that 40% of respondents perceived a trend toward reshoring to the United States within their industries.[9] More than 60% of respondents indicated that the stability of transportation costs would become more important in their location choices in the next 3 years. Additional logistics-related factors were expected to become more important, including the availability of knowledgeable logistics service providers, the availability of transportation, and the reliability of transportation.

Companies are re-examining where their manufacturers and suppliers are located in an effort to expand beyond cost, considering risk and total cost in its place. The cost of moving to a different geography is an important consideration in choosing a manufacturing and outsourcing location. The decision is being approached with a longer time horizon. Companies are considering more strategic issues, assessing and monitoring competition, and listening to the voice of the customer.

In December 2012, Apple announced that it was going to invest $100 million to contract with third parties to make its Mac computers in the United States. America is Apple's largest market, and being geographically close to markets can help supply chain responsiveness. Moving some production out of China will help Apple to diversify sourcing risks (natural events, political events, protectionism, rising wages, etc.). In 2013, General Electric started to manufacture products in the United States, after 50 years of inactivity in the facilities. The move was largely driven by increased oil prices, decreased electricity and natural gas cost, increased wages in China, increased labor productivity, and a change in priorities for American unions.

Summary

Vertical and virtual integration decisions tend to follow a very unpredictable pattern. As the length of the supply chain continues to fluctuate, relationship management, performance measurement, and an in-depth understanding of the total value-creation process will

---

[8]L. Ellram, W. Tate, and K. Petersen, "Offshoring and Reshoring: An Update on the Manufacturing Location Decision," *Journal of Supply Chain Management* 49, no. 2 (2013): 14–22.

[9]Ellram, Tate, and Petersen, "Offshoring and Reshoring."

become paramount capabilities for success. Value can only be delivered if products and services are in the right place, at the right time, at the right price. This is one trend that may reverse itself in the future.

Information Hoarding to Information Sharing and Visibility

By Randy Bradley, Ph.D., at University of Tennessee.

Why Is This a Game-Changing Trend?

The use of business intelligence and analytics to analyze "big data" and make its key messages visible is a trend that is sweeping across industry and the academic community. A few firms are changing the game by linking together masses of information from multiple sources and then analyzing those data with increasingly powerful hardware systems and business analytics expertise. Firms need to make sure they have a plan to catch this game-changing wave.

P&G is a clear leader in this area. As former CEO Bob McDonald said, "We see business intelligence as a key way to drive innovation. To do this we must move business intelligence from the periphery to the center of how business gets done." The challenge with big data is that there is too much of them to process manually. Companies are overwhelmed with the data. We are seeing that real benefit is being provided by automated systems to sort through masses of data, extract valuable information, systematically feed it into supply chain management systems, and alert employees when human attention is required.

Moderate Progress

The management of information in today's rapidly evolving global environment is key to supply chain excellence. What will it take to get to the next level in information management? Before answering that, let us reflect for a moment on the progress made.

The 2000 study predicted that the information-hoarding to information-sharing trend was one of the least advanced of the 10 game-changing trends, achieving a 3.5 level of maturity on a scale of 1 to 10 (10 being total adoption and 1 representing no meaningful acceptance). Although there was no prediction for 2010 that organizations would be operating close to the total adoption level (i.e., a score of 10), significant progress was expected (e.g., level 8–9 maturity). The research supporting this chapter suggests that industry has made moderate progress in moving toward total adoption of an information-sharing philosophy.

Most responding organizations have yet to achieve total maturity, and some have barely eclipsed level 5 maturity. There remains great room for improvement, as is indicated in Figure 1.12. For instance, the average among the firms responding to the 2012 survey was 7.2 with a range, at one standard deviation, of 5.2 to 9.1. As you can see, virtually none of the responding organizations have reached the point of total adoption, whereas an inordinate number of organizations have been slow to embrace the sharing of strategic and tactical information with suppliers and customers to improve performance across the supply chain. With an average maturity level of 7.2, information hoarding to information sharing remains one of the greatest areas of opportunity to improve supply chain performance.

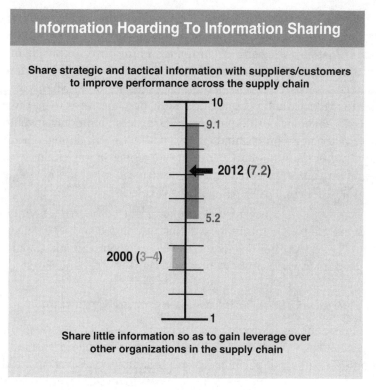

Figure 1.12 Supply chains are shifting to sharing information

Enterprise Systems

The term *enterprise* is used loosely today, as the boundaries of the word have changed over time. Yet many so-called enterprise solutions adopted by organizations have failed to evolve in a manner that is representative of organizational boundary expansions. We must remember that enterprise systems are primarily and generically designed to address the issue of information fragmentation in organizations. On the other hand, "best-of-breed" systems provide customers with software designed for cross-enterprise planning and execution, and a built-in network of trading partners.

Most companies start with their enterprise systems as the first step to harnessing big data. By using the data, companies can sense and respond to quickly changing market realities. It is a game-changer for manufacturers: the result (quantified in Terra Technology's Forecasting Benchmark Study) is an average 40% reduction in forecast error.

How Do We Move Forward From Here?

Moving forward, there are a number of actions that can help move organizations get closer to full adoption of sharing strategic and tactical information with partners/customers to improve supply chain performance. *The secret is an ability and willingness to collaborate with partners and customers.* This is easier said than done when you consider the velocity and volume of data creation and turns (e.g., the continuously changing positions of forecasts, orders, shipments, inventory). This challenge is complicated enough within the traditional enterprise and is rather daunting in the context of a global supply chain network, with multiple tiers of partners trying to manage information exchanges across a variety of hardware and software platforms.

The systematic use of downstream data to improve operational decisions will be the future for creating responsive and resilient supply chains. Leaders using downstream data are seeing an additional 35% reduction in forecast error according to the Terra Technology study.

The Supply Chain Network As a Social Organism

Consider that a supply chain network is a social organism whose success/value is predicated upon its ability to enable or support collaboration among all participants and stakeholders. Given that today's global supply chains require reliable access to real-time, cross-network data, high-quality information begins with a solid, scalable integration platform that connects all trading partners in the extended network. The form of collaboration needed goes beyond the traditional approach of one-to-one sharing of documents. Rather, organizations will need to achieve 360-degree level visibility based on real-time information across a network that provides a single source of truth.

This requires "one-to-many" and "many-to-many" sharing of data and information: one organization shares with many partners/customers (as necessary), and all those partners/customers share with their extended partners/customers. In essence, this approach enables all relevant participants—within the organization and across the supply chain network—to access a shared version of the truth in real time. Additionally, this approach to information sharing provides more than just access to meaningful data; it also provides access to a variety of smart people, all working together with timely, accurate data across the global network. The end result will likely be faster, better decisions that can yield improved cost-efficiency, profit, partner relationships, and customer satisfaction.

In line with this trend, most major retailers in North America have started sharing data with partners (including BJ, Costco, CVS, Dollar General, Family Dollar, Food Lion, HEB, Home Depot, Kmart, Kroger, Lowes, Meijer, Petco, Petsmart, Publix, RiteAid, Safeway, Sam's Club, SuperValue, Target, Walgreens, and Walmart). The challenge for manufacturers is to use the latest technology to make meaningful change.

Emerging Issues in Information Sharing

Finally, three relatively new and emerging issues must be addressed in your quest to reach total adoption of information sharing.

Big Data and Key Insights

Avoid getting caught in the trap of focusing on and being inundated with big data. Remember that value resides in the insights (transformational information derived from the data) that can be leveraged for improved competitive advantage. Focus on those insights.

Data Quality

Determine what constitutes high-quality data by developing specific metrics for measuring quality of the information. Data are only good if they yield good, actionable insights. As such, the tools you use to store, aggregate, and analyze the data play a vital role in determining the quality of your data. Remember, all analytics tools are not equally adept at handling certain types of data and their inherent anomalies.

The Breadth of Your Anticipated Supply Chain Network

Choose solutions that enable or enhance your ability to improve collaboration among members of your anticipated supply chain network. Consider what the breadth of your supply chain network (e.g., multi-echelon, domestic, global, multi-party) will look like and recognize that it, in essence, should be a starting point for determining your "new enterprise" (i.e., your new boundaries).

Summary

In summary, the topic of information sharing is not simply an enterprise system—addressable area in the traditional sense. Rather, this sharing approach and getting results from it require organizations to focus on integrated enterprise solutions that provide customers with software designed for cross-enterprise planning and execution as well as a built-in network of trading partners. They must focus less on so-called enterprise solutions designed only for inside the traditional enterprise (i.e., inside their own organizational boundaries). In essence, organizations that want to reach total adoption of information sharing must embrace the idea that this is as much a network issue as it is an enterprise system issue. This is especially true since the network is the glue that ties trading partners and their systems together to collaborate in a manner that can lead to improved performance in the supply chain.

Managerial Accounting to Value-Based Management—Using Supply Chain Excellence

By Paul Dittmann, Ph.D., Executive Director,
The Global Supply Chain Institute

Why Is This a Game-Changing Trend?

The key message in *The New Supply Chain Agenda* is that the most neglected pathway to increasing shareholder value runs through supply chain excellence. To change the game, firms need to leverage the full potential of their supply chain to achieve breakthrough financial performance.

For example, in the 1990s, the relationship between supply chain excellence and shareholder value was not well understood. Gary Balter, Managing Director of Credit Suisse, observed that few analysts likely appreciated the major change that occurred at Target in the late 1990s and early 2000s. They went from a distribution system clogged with slow-turning merchandise to a flow-through system, with distribution centers dedicated to carrying fast-turning merchandise. Balter observed that this resulted in a major reduction in inventory, with improved product availability. But as their supply chain improved, so did their relative stock market performance versus Walmart and Kmart.

Interestingly, when Walmart began their Remix supply chain program later, all stock market analysts focused on it. Walmart highlighted it because by that time analysts and Wall Street were beginning to appreciate the positive impact and importance of supply chain. More CEOs and board members are taking notice due to stock analysts' consistent questions regarding the state of the firm's supply chain, as well as Wall Street's reward for supply chain performance.

Managing Overall Company Value With the Supply Chain

More and more firms clearly understand that their supply chain is perhaps the most critical lever in creating shareholder value. What will it take to get to the next level in optimizing the overall value of the firm through supply chain excellence? Before answering that, let us reflect for a moment on the progress made.

In our survey of over 160 firms, firms have made much progress in the past 13 years in their commitment to measure individual department performance (based on the overall value delivered to the firm vs. budget-focused, functionally specific metrics). The Managerial to Value-Based Management (Figure 1.13) chart shows much progress, with more opportunity ahead.

The opportunity to improve in this area should not be underestimated. In fact, many firms need to expand their thinking dramatically so that it considers this type of management. The supply chain can be the prime driver of shareholder value (or owner's equity) in firms, yet most firms have not fully leveraged this pathway to making that the case.

Driving Shareholder Value With Your Supply Chain

Given the hype of the last 10 years surrounding the supply chain excellence of companies like Walmart, Toyota, and Amazon, why do

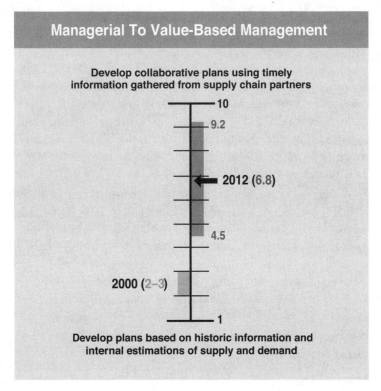

Figure 1.13 Focus on shareholder value is improving the quality of SC decisions

so many firms still not get it? The success of firms in Gartner's Top 25-ranked supply chains, such as Apple, IBM, and Procter & Gamble, should have focused everyone on supply chain as the driver of shareholder value. We hear a lot of talk about the importance of supply chain, but actions often do not match words.

The most neglected pathway to increasing shareholder value (or owner's equity in privately held companies) runs through supply chain excellence. This is not a cost-cutting argument—though supply chain excellence often dramatically reduces cost over the long term. In fact, reaching supply chain excellence is expensive, both in terms of executive attention and actual cash outlays. Supply chain excellence drives shareholder value because it controls the heartbeat of the firm—that is, the fundamental flow of materials and information from suppliers through the firm to its customers. Unfortunately, too many companies have a supply chain where a lack of strategy, lack of talent, a misapplication of technology, internal and external silos, and a basic lack of discipline in managing change cripple this flow.

The supply chain is not just trucks, pallets, and warehouses. But being trapped in that traditional view is one of the primary reasons that few companies are taking advantage of the shareholder value opportunity presented by supply chain excellence. Many executives we talk to are skeptical that investing in this new, expansive vision of a supply chain is worth it. So we will begin by looking at the unequivocal link between supply chain excellence and shareholder value by focusing first on economic profit, which is the linchpin between the two.

Driving Shareholder Value With Your Supply Chain by Creating Economic Profit

Economic profit is profit less the cost of capital needed to generate that profit. Economic profit is important because it means the company is delivering returns above the cost of the capital invested. Figure 1.14 shows the numerous inputs to economic profit and shareholder value. Generating economic profit should be the prime goal of all firms. Most CEOs intuitively know that economic profit drives shareholder value. But many do not clearly comprehend the link that begins at supply chain excellence and continues to shareholder value via economic profit. Supply chain excellence can deliver the most upside to economic profit and shareholder value because its full potential has been so underutilized in the past versus other corporate initiatives.

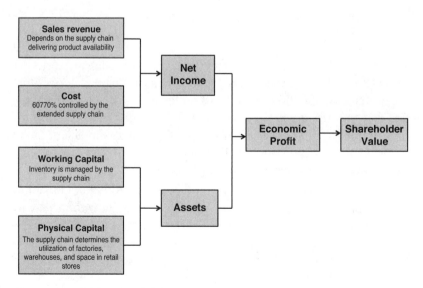

Figure 1.14 Driving shareholder value requires multi-functional excellence

Increased Economic Profit Means Increased Shareholder Value

When economic profit increases over time, shareholder value increases. Stern Stewart & Co. has done extensive research on this concept, which they call EVA (economic value added). They have shown (through extensive analysis of many companies) that the relationship is very strong, especially over time and when the data are normalized.

The Supply Chain Drives Economic Profit

In an increasing (but still small) number of firms, the CEO and the board understand the value of the supply chain to their firm. But many other CEOs, battered by an immense range of items competing for their attention, do not clearly see this link. In many firms, the supply chain controls most of the inventory, manages 60% to 70% of the cost, provides the foundation to generate revenue by delivering outstanding product availability, and manages most of the physical assets of the firm.

The Great Recession of 2008 to 2010 dramatically increased the focus on economic profit. In an era of tighter credit, supply chain levers can be used to free cash reserves from balance sheets rather than depending on restricted credit markets. The opportunity to increase shareholder value in the future will be to take care of both the income statement and balance sheet through supply chain excellence.

Summary

Today, a small but growing number of companies are reporting that they leverage their supply chains to make working capital and cash flow improvements that drive economic profit and shareholder value. Future supply chain organizations must focus on far more than just driving out costs and improving product availability. Instead, they need to become an engine of overall financial improvement for their companies. Smart companies will use innovations in their supply chains to generate the cash to fund innovations in their product lines and growth in their business.

When the credit markets froze in the period 2008 to 2009, a few firms realized that they could free up cash internally without having to go to the banks. A study by Ernst & Young showed that $1.2 trillion (equivalent to 7% of their aggregate sales) is unnecessarily tied up in working capital across 2,000 of the largest companies headquartered in the United States and Europe.

A major lesson learned in our work with participating firms is that *this focus must be driven from the top of the company*. Without strong, consistent support by the CEO, the CFO, and the COO, many supply chain initiatives cannot be successful due to the massive alignment of functional silos required. The fundamental learning from the case, surprising in its power, was how supply chain can be used as a lever to dramatically lower working capital and improve cash flow. Since these changes positively affect economic profit, investors reward these efforts as they realize higher shareholder value.

The supply chain will not drive economic profit without a supply chain strategy. After working with hundreds of firms, we have surprisingly found very few that have a real supply chain strategy. To succeed, a very basic prerequisite exists. The supply chain organization must challenge itself to take a broad economic profit–based view, and they must understand this language and the language of the board.

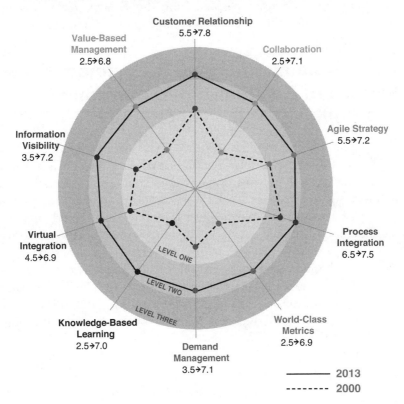

GAME-CHANGING TRENDS EVOLUTION

Chapter 1 Addendum: Game Changers Update—10 + 5 Trends for 2025

As the Game-Changers research showed, the environment for supply chain activities has changed considerably in the first 15 years of the 21st century. Yet, while those changes were significant, many experts predict that even more substantial change will occur over the *next* decade. In fact, the supply chain world of 2025 promises to look very different from the one today.

In 2015, Ted Stank and Chad Autry of the University of Tennessee and Pat Daugherty and Dave Closs of Michigan State University published an article in *Transportation Journal* that extrapolated on the state of the original 10 game-changing trends in order to predict the evolution of supply chain management over the coming decade.[10] The authors also identified five new trends that are likely to be influential in driving the changing supply chain between 2015 and 2025. The following sections, excerpted from the *Transportation Journal* article, predict the state of each game-changing trend in 2025.

Game-Changing Trend 1: Customer Service to Relationship Management to Intelligent Value Co-Creation

A nuanced shift in focus from reactive customer analysis toward intelligent value co-creation is envisioned over the next decade. Firms are already purposefully using analytics to conduct segmentation and optimize segment service levels. We expect this process to accelerate by way of automation. Specifically, we predict a continued emphasis on value optimization through the use of system automations that will enable organizations to offer multiple, different, and ideal service levels within the same end-to-end supply chains and horizontally across functional business units. The emergence of "big data" and "small data" extraction tools and intelligent heuristics/programming should allow firms to integrate business rules surrounding service offerings,

---

[10]T. Stank, C. Autry, P. Daugherty, and D. Closs, "Re-Imagining the Ten Mega Trends That Will Revolutionize Supply Chain Logistics," *Transportation Journal* 54, no. 1 (2015): 7–32.

both internally and across partner interfaces. This insight will help managers determine the ideal service level for a given multi-echelon chain based on differences in only a single intermediate and/or end customer node, thereby maximizing returns on cost-to-serve for the focal firm in the supply chain. In summary, as service stratification is pushed out within the firm and through its partners, the future will bring service strategies that are fitted to product-market situations under something resembling a "mass customization" logic, in great contrast to the one-size-fits-all supply chains of a decade ago.

Game-Changing Trend 2: Functional Focus to Process Focus to Systemic Focus

We believe that there will be significantly more focus on technology-enabled supply chain integration on the horizon. Supported by greater visibility, processing capabilities, and collaborative capabilities afforded by emerging computer and cloud technologies, firms are beginning to optimize process flows, not simply within their own operations but also between their processes and those of their partners. It is not farfetched to expect that this type of integration could soon synthesize multiple processes spanning the entire supply chain. By focusing on optimizing the entire supply chain network (inclusive of all processes subsumed within it) rather than focusing on individual functional linkages nested within smaller network segments, the supply chain itself becomes solution-oriented rather than process/flow-oriented. The resulting integrated supply chain could then be measured by only a few unified—but meaningful—metrics, and could be driven almost entirely by a single forecast. These opportunities will only grow as information accuracy and governance continue to improve over the coming years due to simultaneously enhanced enterprise resource planning and execution system connectivity across multiple organizations.

Game-Changing Trend 3: Incremental Change to Transformational Agility to Prognostic Agility

Previous research has demonstrated that firms still have much progress to make toward achieving adaptive capabilities that can transform the competitive environment. Challenges to supply chain agility—including changing geopolitical events, demand patterns, supply locations, distribution patterns and trends, and technological innovations (such as the "Internet of Things")—continuously render existing infrastructure and practices obsolete. Rather than reacting to patterns that are largely unpredictable, firms must increasingly achieve *prognostic agility* over the next 10 years. Prognostic agility enables supply chain decision-making processes that proactively account for both existing and anticipated competitive situations. This will require expanded total landed cost-to-serve models that include risk, sustainability, and social costs. Continuous and dynamic situation analysis and modeling will facilitate such models. It will also enable firms to proactively create different supply chain solutions for different product, customer, and environmental segments.

Game-Changing Trend 4: Information Hoarding to Information Sharing to Information Synthesis

Many organizations have been slow to fully embrace the value of sharing strategic and tactical information with suppliers and customers. What must firms do to fully realize the potential of information sharing and take the next steps to information synthesis? An initial step is to fully develop collaborative arrangements with partners and customers. This will not be easy—particularly in the realm of global supply chain networks, where multiple tiers of partners deal with a variety of hardware and software platforms. Sharing proprietary information and unlocking the value of big data offers immense potential for developing business intelligence and exploiting analytic capabilities. It can allow examination of the total supply chain network of partners in contrast to the traditional enterprise approach of only examining operations within the organization. While few firms at this time can link masses of information from multiple sources and analyze those data to achieve competitive advantage, this capability is expected to increase. More firms are focusing on acquiring or developing the necessary talent to exploit data for competitive advantage. The firms that do learn to holistically share and jointly interpret information will enhance their competitive stature. They will, in effect, be creating new supply chain insights regarding customer, product, supplier, and operational interdependencies.

Game-Changing Trend 5: Adversarial Relationships to Collaborative Relationships to Vested Relationships

Over the next decade, it is likely that successful collaborative relationships will move to a new level of cohesion that will be very difficult to replicate. A greater proportion of trading partners will work to develop "vested" relationships. In a vested relationship, the parties build a mutually equitized relationship based on shared values and goals. The working arrangement is carefully designed such that neither party can succeed without the other succeeding as well. A vested relationship is in direct contrast to adversarial relationships that were so common at one time and still exist today. In an adversarial relationship, there is typically a power imbalance that results in a win-lose situation. The trading partner with the power makes the decisions and can negotiate or dictate the terms.

In contrast, a vested relationship represents the potential for a win-win situation. The focus is ensuring long-term success for both parties by developing operational and strategic plans detailing joint rewards. Vested relationships also require joint accountability if they are to be successful. The ultimate goal is to optimize total system value creation.

Game-Changing Trend 6: Demand Forecasting to Demand Endcasting to Demand Shaping

One discovery noted in the previous research is that there is still opportunity for greater maturity in supply chain partners' collaborative forecasting. Firms must construct forecasting and demand planning processes that "sense and respond" to dynamic and evolving customer/consumer behaviors. This can be done by mining data from multiple sources, including social media, public blogs, and relevant subscription-based services. The data must then be deployed to *predict* the sale of goods to customers at the right time and location (i.e., the demand point). This predictive demand signal should be transmitted through planning and decision support processes (such as Integrated Sales and Operations Planning) and then through the actual execution from Sales Order Processing to manufacturing execution.

Game-Changing Trend 7: Training to Knowledge-Based Development to Talent Management

The focus of training is moving away from developing functional skills to a broader focus aimed at helping employees develop skills that directly improve overall organizational performance. Firms have also recognized that a more appropriate starting point is to acquire the right talent to fit the skill and competency profiles needed. The narrow, functional training perspective has commonly been replaced with a broader-based focus of finding, hiring, developing, and retaining talent. This broadened perspective is expected to continue in the future. Firms will focus on talent management and are likely to take a much more proactive approach. Retirements of the baby boomer generation will continue to create shortages that are likely to escalate further in the future. Firms realize they must do succession planning to ensure employees are ready to step into these vacated positions. However, managers are likely to take a more structured and thoughtful approach than they may have in the past. They will spend more time immediately identifying the critical skill needs and how those needs are likely to change in the future.

There is also likely to be greater emphasis placed on ongoing training and development of employees once they are hired. It is becoming much more common for firms to commit to in-house and external training and educational programs. Talent retention is going to be even more critical. Today's workforce is unique; generational differences require managers to reevaluate traditional work structures and incentive/compensation programs. Flexible approaches may be the answer to retaining top employees and keeping them committed and actively involved.

Game-Changing Trend 8: Vertical Integration to Virtual Integration to Flexible Network Integration

Creating appropriate structures and processes to facilitate virtual integration has proven to be far more challenging than anticipated. Over the next 10 years, improvements in virtual integration governance structures and processes—plus technology innovations (such as cloud computing and nonproximal additive manufacturing)—will serve to greatly reduce the risks associated with virtual integration. As a result, firms will increasingly select and engage a dynamic constellation of partners to perform upstream and downstream supply chain activities to optimize total system value depending upon prevailing requirements of value creation. As global business potentially decentralizes into regional clusters of demand and supply, the resulting complexity will require collaboration with multiple outsourcing partners to maximize total value to customers.

Game-Changing Trend 9: Functional Measurement to Customer Service Measurement to Relevant Value Measurement

Over the next 10 years, the best firms are likely to develop even more relevant, targeted measures for use in assessing the value they are providing their choice customers. Currently, data are readily available from a range of sources. The key is to determine the critical data, the right metrics, and the appropriate goals. It is not enough to focus on internal firm success. A longer term and external focus will also be required. Consideration must also be given to supporting customers so that their success continues as well.

Game-Changing Trend 10: Managerial Accounting to Value-Based Management to Total Value Orientation

While there has been a growing awareness of the potential for the supply chain to be a driver of value for key customers, this is expected to become even more prevalent in the future. These expectations should extend to other stakeholders and supply entities. Eventually, this will lead more firms to embrace a total value orientation. Organizational objectives will be driven by the desire to have all involved parties share in the creation of value. Resources in the form of working capital, human resources, and physical assets will be leveraged to drive shareholder value. To be truly successful, these efforts will need to be driven from the top of the company. Strong, consistent, C-level support will be needed.

The "+5" Game-Changing Trends

In addition to the 10 game-changing trends identified within the original Bowersox, Closs, and Stank (2000) research, other forces have emerged that will impact competitive conditions in and around firms and their supply chains.[11] These forces were not on supply chain managers' radar at the time of the original research.

Game-Changing Trend 11: Transformation

Supply chain managers can no longer afford to prioritize and allocate resources purely to enhance input and output stability. Customers for previously local or regional businesses are located worldwide, use the Internet to shop, and live in areas formerly considered unviable. Business with these types of customers is made possible because demand centers possess more disposable income than ever before. Accordingly, a substantial portion of resource investment must now be allocated to driving desired changes in outcomes in response to dynamic competitive conditions. Although many firms are starting to spend money on supply chain innovation in order to meet unpredictable and widely distributed demand, most innovations continue to be focused on products rather than supply chain processes, in spite of the fact that current supply chain processes are poorly designed for nascent offerings.

Many examples of supply chain innovation are emerging, and we expect this trend to become even more transformational. We believe that firms will increasingly realize the value of prioritizing and assigning resources to do two things: address current supply chain issues due to demand/supply dynamics (i.e., the current drivers of current supply chain innovation) and prioritize and transform resources in ways that shape future outcomes, in advance of competitive conditions.

This shift from supply chain innovation to transformation should occur not only in product-focused contexts (as in the case of Amazon's drones experiments, or the numerous advantages of additive manufacturing already on display), but also in intangible/service-based supply chains. For example, ordering via the "Internet of Things" makes it possible for auto repair services (and the procurement or

---

[11]D. Bowersox, D. Closs, and T. Stank, "Ten Mega Trends That Will Revolutionize Supply Chain Logistics," *Journal of Business Logistics* 21, no. 2 (2010): 1–15.

manufacture of associated parts) to be initiated when a vehicle's internal sensor system detects a problem and transmits a distress signal to the dealership. In this case, the customer would not have to schedule a repair visit. Similarly, a customer's superstore trip to pick up laundry detergent will be initiated by her washing machine, and executed via a shared delivery outsourced by the grocery chain (think Uber for groceries) rather than a detour on her way home.[12] Such process transformations will necessarily be accompanied by a reimagining of supply chain relationships that are designed around flexibility and problem solving in real time rather than fixed contracts and cost management.

Game-Changing Trend 12: "Dumb" Technology to "Smart" Technology to Autonetic Technology

Many of the future recommendations that stem from the game-changing trends we have presented thus far are fundamentally influenced by technological changes. However, technology itself—especially in the supply chain space—is transforming the lens through which managers view problems. It is for this reason that technology warrants independent consideration. It is easy to forget that Internet-enabled E-commerce has been viable for less than 20 years. Prior to our original research, supply chain information systems were frequently "dumb" terminals and electronic data interchange (EDI) connections. These applications were effective at accepting, transmitting, and displaying data. But they possessed limited capability for automated processing and "intelligently" supporting decision making, especially when partners were operating different platforms and/or were geographically separated. Many companies transformed their supply chain operations during the subsequent decade by interjecting process-enhancing technologies that augmented the data portability of "dumb" technologies with "smarter" applications that added knowledge value, as in the cases of Smart RFID, container scanners, etc., that proactively broadcast information regarding the products and shipments they are tracking. However, many firms have been reluctant to adopt such technologies due to start-up costs and turbulence in the technology industry.

---

[12]K. O'Marah, "Eight Disruptive Technologies Impacting Supply Chain," http://www.scmworld.com/

As cost and confusion surrounding smart supply chain applications and hardware continue to decrease, we expect that many more firms will adopt and implement these in the near future. But we also believe that the turbulence will continue in the form of autonetic technologies. Although smart technologies can currently process supply chain data to support (close to) real-time decision making and problem solving, *autonetic technologies* combine this capability with predictive data analytics in order to anticipate—and alleviate—supply chain breakdowns before they occur, based on probability distributions.

Consider that both Kimberly-Clark and Amazon are already shipping goods in the *general direction* of customer locations based purely on demand signals instead of actual point-of-sale data. Final diversion to the actual customer location occurs only when the customer has actually placed an order. These predictive technologies, in combination with further diversified AI-based applications that "learn" from supply chain mistakes and adapt networks in real time, will be a key differentiator in 2025. We expect their adoption to be less than ubiquitous, however, due to expenses associated with big data identification, merging, and reduction (back to small data) for analysis. Many firms will probably lag in adopting autonetic technologies.

Game-Changing Trend 13: Local Optimization to Global Optimization to "Glocal" Optimization

The 2000s represented a time of unprecedented global supply chain centralization. Firms were able to take advantage of the open regulatory and tax environments. They applied complex analytical models to determine the most efficient locations for global supply and manufacturing by optimizing tradeoffs between land, labor, manufacturing, facility, and transportation costs. Yet challenges to the global model have emerged during this time as well. These challenges are a result of the dramatic increases in supply chain complexity, risk, and rapidly shifting cost curves associated with labor and transportation.

The Great Recession of 2008 to 2010 motivated many companies to reevaluate their global supply chain strategies. Beyond the potential for higher total costs associated with global supply chains, many firms also cite risks emerging from intellectual property theft, product quality problems, and limited access to capable third-party service providers as reasons for challenging the centralized global supply chain model. Other factors influencing global network optimization decisions include:

- Increased supervision and training
- Higher local security
- Extra expenses associated with travel and telecommunications

As a result, firms are increasingly considering a regional model for servicing demand. In regional models, supply and manufacturing networks are located relatively close to large demand markets. While opportunities for centralized global operations for certain supply and product segments will continue, a more enlightened perspective on supply chain optimization will arise in the next 10 years. This will result in *glocal optimization*, characterized by a segmentation approach that sees some elements of a supply chain optimized on a global basis and other elements decentralized to regional or local levels.

Game-Changing Trend 14: Risk Agnostic to Risk Management to Risk Prognosis

The September 11 tragedy served as a pivotal point for many businesses to design formal strategies for systematically identifying and addressing supply chain disruptions. It also taught these firms that they must institute mitigation strategies. The notion of risk management became an operational and financial concept. Facility and supplier contingency and crisis plans came online. And supply chain risk portfolios that were tied to economic value for many companies (based on probability and predicted magnitude of disruption occurrence) began to emerge.

Given the constancy and magnitude of changes in the global business environment, we believe that, over the next decade, firms of all sizes, origins, and missions will be forced to deal with increasing risk in their supply chains. We believe that leading SCM firms will move from a risk management focus (emphasizing that risk, once apparent, will be effectively managed) to one of risk prognosis. Risk prognosis is a focus on creating processes that leverage business intelligence to reduce (or eliminate) supply chain risk before it emerges by designing supply chains that exclude risk factors. Ideally, the cost of the various forms of risk will be incorporated in expanded total cost-to-serve models, and supply chain value-creation decisions will be based (at least in part) on the risk profiles calculated for entire expected supply chains. In the future, we expect this technique to be applied vertically to several tiers within the same supply chains. This will allow firms to assess probabilities and costs much closer to—or even at—the raw material source.

Game-Changing Trend 15: Untenability to Sustainability to Prostainability

The notion of sustainability was essentially unheard of at the time of the original research. Most individuals and businesses viewed global resources and markets from a purely "cornucopian" perspective—meaning that long-term impacts of resource shortages or social disruptions were dismissed due to a pervasive availability of substitutable options. At the turn of the century, any concerns of future market inefficiencies due to resource depletion or social unrest/injustice were viewed as zealot theatrics rather than genuine business continuity threats. This was because alternative supply sources (or product redesigns) could be integrated without too much operational disruption. For many firms (especially those headquartered outside Europe, where stiffer regulations prevail), sustainability remained an issue for discussion and analysis—but not yet for action.

Due to the long-term nature of sustainability paybacks, it must become a part of the organizational DNA (vs. simply a company project) before its proposed benefits will be actualized. In the future, this would imply that firms design their supply chains with sustainability outcomes (and identification of the partners who could facilitate their achievement) in mind. However, the firms who will achieve the greatest returns on sustainability will be those who take this notion one step further by becoming "prostainable." It will not be enough in the next 10 to 15 years to simply minimize and/or balance social, environmental, and economic costs. Firms will actually need to design supply chains that proactively seek to improve on relevant outcomes instead of simply sustaining operational viability. Those firms who not only reduce environmental and social damage but also actually improve the status of each, while remaining economically viable, will be the overall winners.

We believe that products such as cars that reduce smog, manufacturing plants that generate more power than they use, or machines that are powered by trash or energy waste while creating viable jobs for local operators will become the leaders of their industry sectors, improve overall operational efficiency, and satisfy the requirements of increasingly socially conscious customers. We expect companies to react begrudgingly at first. Eventually, they should be more enthusiastic to opportunities for sustainable development, especially as it becomes increasingly apparent to business leaders that balancing social and environmental goals with economic objectives provides market benefits and positive financial returns.

Summary

The above narrative highlights the impact on supply chain management emerging from the many changes in the external business environment, including changes in consumer and customer preferences, changes in strategic focus, geopolitical upheaval, and technological innovation. The following list summarizes the expected state of the 10+5 Game-Changing Trends in 2025.

- *Intelligent value co-creation*: Focus on developing automated stratification processes and value-creation delivery.
- *Systemic focus*: Focus on optimizing the entire supply chain network of processes to optimize customer value co-creation.
- *Prognostic agility*: Decision-making processes proactively account for both realized and anticipated competitive situations.
- *Information synthesis*: Information is holistically shared and jointly interpreted to create supply chain knowledge and understanding of how it affects performance.
- *Vested relationships*: Focus on creating joint accountability and rewards to motivate each supply chain partner to optimize total system value creation.
- *Demand-shaping*: Focus on proactively influencing demand to conform to supply chain capabilities that optimize value co-creation.
- *Talent management*: Proactively sourcing and developing talent by identifying the critical range of skills needed for future success and education required to keep employees effective and involved.
- *Flexible network integration*: Selecting and engaging a dynamic constellation of partners to perform upstream and downstream supply chain activities to optimize total system value depending on prevailing requirements.
- *Relevant value measurement*: Measuring operational performance based on the creation of value for customers of choice and supply chain entities.

- *Total value orientation*: Driving organizational objectives based on the pursuit of value maximization for both customers of choice and supply chain entities.
- *Transformation*: Prioritizing and assigning resources to shape future outcomes in advance of competitive conditions.
- *Autonetic technology*: Prevalence of technology that anticipates a range of potential data inputs and proactively devises solutions based on probable likelihoods.
- *Glocal optimization*: Positioning supply chain activities and processes to optimize total system performance across multiple demand centers.
- *Risk prognosis*: Instituting formalized processes to reduce or eliminate supply chain risk prior to its occurrence based on business intelligence.
- *Prostainability*: Designing supply chains that actively seek to improve social, environmental, and economic outcomes.

Transition: Game-Changing Trends in Supply Chain to Global Supply Chains

The past 15 years have seen dramatic shifts in supply chain strategy. These game-changing trends emerged in the 2000 landmark study, and companies have been making steady progress toward total adaptation ever since. In that same period of time, globalization has transformed from an impending reality to an ever-present one.

As businesses become global, their supply chains naturally follow suit. The two-part nature of game-changing trends just so happens to characterize global supply chains as well. When firms expand their operations around the world, profit and value are exponentially affected. Exchange rates, tariffs, and transportation variation all have an effect on the bottom line. Global supply chains are not exactly simple to implement, either. International sourcing becomes incredibly difficult to accomplish when the supply chain becomes intercontinental.

When taken to a global scale, each game-changing supply chain trend is greatly exacerbated. Consider the "Adversarial to Collaborative Relationships" trend, for instance. Committing to collaboration, aligning goals, and sharing information are difficult enough for supply chains without language and culture barriers. Introducing distance and dissimilarity can only make things harder.

Chapter 2 takes a global perspective on supply chain management. The section begins with a brief history of the firm in a global context and the gradual shift to a TCO approach. After the foundation for global supply chain management is established, we present the EPIC framework as a means of assessing supply chain readiness across global regions.

After dissecting the EPIC framework and how it can help managers assess global opportunity, we detail two best practices that best-in-class companies follow when they make global sourcing decisions. The first—supply chain network design—has five components that drive the behavior behind global sourcing analysis and decision making. The second—managing complex, global supply chains—also has five components to success. Within each of these practices are specific examples of how major industry players harnessed (or failed to harness) the recommended strategies.

Global supply chain management has always been a dynamic field. The game-changing trends discussed in Chapter 1 are only magnified as they are applied on an international stage. To effectively direct a global supply chain, managers must be that much more mindful of the game-changing trends. They must also be prepared to benchmark themselves against elite competition, using the best practices presented here as their guide.

Global Supply Chains

The Fourth in the Game-Changers Series of University of Tennessee Supply Chain Management White Papers

Ted Stank, Ph.D., Mike Burnette, Paul Dittmann, Ph.D.

Executive Summary: Best Practices for Managing Global Supply Chains

Over the last several decades, business has become truly global. Although globalization has benefitted many organizations via the creation of new markets, it also has presented serious new challenges for supply chain executives who often struggle to achieve desired customer service, quality, cash, cost, responsiveness, and innovation standards. The University of Tennessee's Global Supply Chain Institute (GSCI), in cooperation with sponsor BT Global, has developed the following chapter to capture best practices for supply chain leaders seeking to design and manage their global supply chains.

The paper begins with insights from a recent book authored by a team of faculty from the University of Tennessee and ESSEC School of Management (Paris, France) entitled *Global Supply Chains: Evaluating Regions on an EPIC Framework*. The EPIC framework originally introduced in the book provides a methodology and approach that enables firms to better assess different locations for locating global supply chain operations, including sourcing, manufacturing, and distribution. To validate the premises of EPIC, we interviewed 10 companies that we believe possess best-in-class global supply chains in order to glean best practices. These interviews are also described in the chapter.

Global supply chain challenges are not new. However, supply chains over the last 15 years have been particularly dynamic with the emergence of economic opportunity across South and Southeast Asia, as well as in select locations in Eastern Europe, South America, and Africa. The rapid ascent of developing nations within these regions—China and India in particular—held the promise for more than a decade of reaching pockets of intellectual and financial capital from across the world, along with major cost reduction. As a result, many firms had offshored much of their product sourcing and manufacturing to Chinese or other Asian suppliers by the mid-2000s.

This rush to Asia was often driven by cost-cutting strategy. In general, when organizations pursue cost cutting without giving much thought to associated impacts on customer service, they operate in a *cost world*, and not a *total cost of ownership* (TCO) world. By TCO, we mean "full cost accounting," where all conceivable costs (including direct, indirect, and the cost of lead time and lost sales) are considered. The adoption of a short-term, cost-cutting mindset inevitably drives firms toward less-than-optimal decisions and strategies that will lead to poor long-term outcomes. A key question addressed in this chapter is: "Why did so many organizations find that the expected savings from global sourcing initiatives failed to materialize?" An obvious answer is that these organizations simply did not do their homework. They focused on short-term cost *reductions* without considering all of the hidden costs associated with the *TCO* of their offshoring decisions and neglected to consider the significant potential negative impacts this approach would yield for supply chain responsiveness.

Our research demonstrates that only a very small percentage of organizations fully consider many of the hard-to-compute supply chain costs that can severely hurt an organization's competitive advantage. These costs include the cost of lead time, the cost of (in)flexibility, the cost of quality, the cost of lost sales, and of course the costs associated with the added risks that exist within a global environment. Costs such as these give a whole new meaning to the phrase, "Distance matters." Today, the organizations that pay attention to these costs of globalization tend to focus more on near-shoring—or perhaps even reshoring—their supply chain operations.

Additionally, though many organizations will acknowledge that they did not conduct a thorough TCO analysis prior to making global network decisions, they are reluctant to acknowledge a more profound problem: they struggle to adequately execute their supply chain in the dynamic global environment. Going global increases cost, complexity, and risk. Simultaneously, managing these three aspects can be extremely challenging.

The story of the second decade of the 21st century will be different. As global supply chains proliferate, organizations operating in one country will increasingly depend on organizations headquartered or operating within the boundaries of other countries to either supply material or market their products. Our research suggests that global supply chains across the world will eventually break into a series of demand and supply pods where regional procurement and manufacturing operations will supply the major demand centers of the area, at least for a significant percentage of production requirements.

Supply chain professionals who operate in the global environment need to be armed with a solid global supply chain management strategy. An important tool needed to develop this global supply chain strategy is the EPIC framework from *Global Supply Chains: Evaluating Regions on an EPIC Framework*.

Using the EPIC Framework to Evaluate Global Regions

Global supply chain managers can benefit from a tool that helps them assess their supply chain location decisions, identifying the strengths, weaknesses, opportunities, and threats of the different regions in the world. The EPIC framework provides the structure for assessing various regions around the globe for supply chain readiness from economic (E), political (P), infrastructural (I), and competence (C) perspectives.

The EPIC framework defines and explains these dimensions of the global market environment to assess their potential impacts on the effectiveness of global supply chain management activities and to identify the characteristics of those dimensions in each region of the world. The framework measures and assesses the levels of "maturity" held by a geographic region, with specific respect to its ability to support supply chain activities. The four EPIC dimensions are then assessed using a set of variables associated with each dimension. Each EPIC variable is assessed using a combination of quantitative and qualitative scores based on data drawn from a wide variety of data sources.

Regional assessments for 55 countries are included in the EPIC analysis. The assessments are organized along 10 distinct geographic regions: East Asia, South Asia, Southeast Asia, Australia, Mid-East and North Africa, Sub-Saharan Africa, Western Europe, Central and Eastern Europe, North America, and South America. The assessments utilizing the EPIC framework ranged from D (lowest score) to A. The scores for all 55 countries are listed in the chapter. A summary of the key findings, in the form of narrative themes, follows.

Summary of Key Themes From the EPIC Assessment

- Supply chains have undergone a series of phase transitions over the ages, from trading to manufacturing to the current era of global, IT-enabled supply chains. These transitions have been fueled by political and technological innovations.
- These political and technological innovations resulted in global economic power moving from Asia to Europe near the middle of the 18th century and then to North America following World War II. Recent trends suggest that the balance of economic power is either moving back to Asia or at least being leveled across the Americas, Europe, and Asia. Supply chain professionals must have a robust TCO process as described above to stay ahead of the changes in the dynamic global environment.
- The world is becoming flatter as barriers to free-market capitalism are removed, the use of the Internet becomes more widespread, and goods increasingly flow across borders. However, the extent to which the playing field is being leveled is up for debate. The flattening trend could reverse course. Stranger things have happened.

Best Practices

Equipped with the EPIC framework, we talked with 10 leading supply chain companies that have demonstrated best-in-class capability in global supply chain management. These companies came from consumer packaged goods (CPG), chemical, luxury, materials refining, food, supply chain services, and health care industries.

In the interviews conducted to identify global supply chain best practices, the benchmark supply chain organizations addressed the need to have strong capability in two important areas:

1. Ability to make high-quality supply chain network design (SCND) decisions
2. Systems to manage complex, global supply chains

SCND—Global Sourcing Analysis and Decision Making

In making global sourcing decisions, our best-in-class companies followed five best practices:

1. Supply chain decisions are *strategy driven*
2. These firms strive for *scale*
3. Supply chain organizations have a core competence in *fact-based, quantitative analysis*
4. Supply chain decisions are *net present value* (NPV) and *TCO-*based
5. Supply chain decisions are based on *holistic product designs* appropriate for the global environment

Best Practices for Managing a Complex, Global Supply Chain

In our discussions with the best-in-class supply chains, we also saw five best practices in managing their complex global supply chains:

1. An effective *global Sales and Operations Planning (S&OP)* process
2. A *process to manage complexity*, especially product complexity
3. Strong *supplier collaborative partnerships*
4. A *talented team* on the ground
5. *Clear visibility* throughout their global supply chain and *rapid response capability*

In the chapter, we discuss these best practices along with specific examples of successes and failures to illustrate the learning that drove the best practices.

We wanted to make a special note of the last best practice listed above: "clear visibility and rapid response systems." Managing supply chain risk is an expensive and critical process in global supply chain management. For this best practice, we have included a supply chain visibility case study in the chapter.

Visibility is extremely important in any domestic supply chain, but it reaches another level in the global environment. A domestic supply chain with lead times of days or weeks can easily stretch to multiple months when placed in a global environment. Countless events routinely disrupt a supply chain over such an extensive time span, particularly in emerging third-world economies. A global supply chain faces disruptions resulting from demand spikes, natural disasters, political unrest, strikes, unexpected regulatory issues, port problems, and terrorism. Tools are available to enhance supply chain visibility. *BT Trace* from BT Global is one example of a leading-edge solution. Table 2.1—the Global Supply Chain Network Design (SCND) tool – is also available to help make global network decisions that are based on business and strategic needs.

Table 2.1 The Best Global SC Network Decisions are Based on Business and Strategic Needs

Global Supply Chain Network Design (SCND) Should you have very few global or multiple local suppliers/plants/warehouses?		
FEW/ GLOBAL	Do you have a low total supply chain cost as % of total revenue? ◀◀◀	
	Do you have a high level of technical complexity? Do you need high level skills to produce? ◀◀◀	
	◀◀◀ Do you have complex technology? Is it proprietary (legally protected)?	
	Is your sales/volume distributed evenly across the world? ▶▶▶	**MULTIPLE/ LOCAL**
	Is your logistics cost a high % of your total SC (value) chain cost? ▶▶▶	
	Does your business require a high level of SC responsiveness? ▶▶▶	
	Are your products regionally specific? Does your business require regional specific new product innovation? ▶▶▶	
	Does your business have a high correlation of customer service defects to lost revenue? ▶▶▶	
	Is Inventory Management (managing cash to the lowest levels) of high importance to business? ▶▶▶	

Sources

This chapter is based in part on material from the recent book *Global Supply Chains: Evaluating Regions on an EPIC Framework* by Mandyam Srinivasan, Theodore Stank, Philippe-Pierre Dornier, and Kenneth Petersen.[1] In addition, new research based on in-depth interviews from 10 leading companies conducted in support of this chapter identifies best practices used to manage global supply chains.

---

[1]M. Srinivasan, T. Stank, P. Dornier, and K. Petersen, *Global Supply Chains: Evaluating Regions on an EPIC Framework—Economy, Politics, Infrastructure, and Competence* (New York, NY: McGraw-Hill Professional, 2014).

Global Supply Chains: Evaluating Regions on an EPIC Framework—Economy, Politics, Infrastructure, and Competence is the outcome of a unique partnership among four coauthors who grew up in different parts of the world: the United States, France, and India. The goal of this book is to provide information about supply chains in each region in the world, identifying their unique characteristics to help the decision maker arrive at more informed decisions. Managers of global supply chains can use the framework developed in this book to help them assess their supply chain strategies, identifying the strengths, weaknesses, opportunities, and threats of the different regions in the world.

Most managers do not have the luxury of spending time "in country" to learn the nuances of global supply chain issues prior to making decisions. Having a handy reference to information critical to good global supply chain decision making can significantly help these executives manage supply chains in both emerging and mature markets. It is the pursuit of such knowledge that has driven the authors of this book to develop the EPIC framework—a structure to assess countries around the globe on their supply chain readiness from four different perspectives. The four different perspectives are economy (E), politics (P), infrastructure (I), and competence (C).

The EPIC framework is used in the book to assess 55 countries in 10 geographic regions around the globe. The framework is intended to help organizations that wish to invest in or manage supply chains in these regions or countries. Each of these dimensions evaluates a number of variables to arrive at a weighted score for that dimension. In turn, the scores on these dimensions are used to arrive at a weighted score for the country.

The research conducted for this book reveals that key variables in the macroenvironment can help managers better understand the framework for decision-making and reduce uncertainty. Knowledge of the levels of these variables enables supply chain managers to choose the locations for value-added supply chain operations for their enterprise, including transportation hubs and modes for raw materials, location of parts and subcomponent suppliers, finished goods manufacturing and assembly locations, and transportation and distribution hubs for finished goods. In particular, the research results reveal interesting combinations of sourcing, manufacturing, and logistics options for different regional consumer markets.

Introduction: Global Supply Chain

Back to the Future

Perhaps you decided to read this chapter because you have a global supply chain and you urgently need to improve it. You most likely have global suppliers and/or global customers. You know you need to manage this complex network to provide better service to your end customers, but you also must do that with ever-lower costs and inventory levels. Our goal is to provide a series of best practice recommendations based on the EPIC framework from the recent book *Global Supply Chains: Evaluating Regions on an EPIC Framework* to help you meet these daunting challenges.

Although the term "supply chain" was not commonplace until the late 20th century, global supply chain challenges are not new; rather, they have existed for centuries. The Phoenicians traded actively with Egypt more than 4,500 years ago, establishing the Middle East as one of the first major crossroads of global trade. From 206 BC into the 15th century, the Silk Road played a significant role in the economic development of China, India, Persia, and Arabia. Asia was the center of supply chain activity until the Industrial Revolution in England dramatically shifted this to Europe.

The United States emerged on the global scene in the late 1800s, fueled by the 1869 completion of a continuous railroad line that stretched across the continent from the Atlantic Ocean to the Pacific Ocean. Soon, trains were carrying freight loaded with cargo such as teas, silks, and other handcrafted items from Japan, India, and China. Spices, fruits, cattle, sheep, and minerals were transported across the continental United States.

Japan reemerged from the destruction of World War II as a global force in the 1960s. The Japanese automotive industry grew from producing just over half a million vehicles in 1960 to producing more than 11 million vehicles in 1980—a number that exceeded US auto production by more than three million vehicles that same year.[2] Japan would remain the world's leading auto producer for the next 10 years.

---

[2]C. Berggren, *Alternatives to Lean Production: Work Organization in the Swedish Auto Industry* (Ithaca, NY: H. R. Press, Cornell University, 1992).

Where It All started: The (Re)emerging Power of Asia

Although Japan may have led the way, the last 15 years have witnessed a reemergence of economic power across Southern and Southeast Asia, as well as in select locations in Eastern Europe, South America, and Africa. While outsourcing to China and other Asian countries may be slowing, the growing economic might of the region and its impact on global supply chains will continue to make it a force with which to reckon. The rapid ascent of these developing nations (China and India in particular) in the early part of the 21st century led *New York Times* columnist Thomas L. Friedman to write his bestselling book *The World is Flat*.[3] Friedman's book posits that a series of events—including the deregulation of trade, liberalization of foreign direct investment (FDI), and the development of the Internet—have combined to "flatten" the world, or level the playing field that the industrialized world has enjoyed since the dawn of the industrial age. As a result of these forces, many Western-based manufacturing organizations found that it was possible to outsource activities, essentially splitting up work and sending individual operational pieces to far-flung locations such as Bangalore and Beijing. The resulting structure enabled activities to be performed virtually around the clock. Such outsourcing also held the promise of reaching pockets of intellectual capital across the world. By the mid-2000s, many firms had moved virtually all product sourcing and manufacturing to Chinese or other Asian suppliers offshore.

While Friedman's vision of the world in the 21st century was initially met with widespread support, it has also been subject to controversy and criticism. His critics argue that the world is far from flat, and that differences between countries remain much larger than acknowledged. Organizations that outsourced to emerging markets initially were enthralled by the dramatic reduction in the cost of goods sold engendered by this new purchasing arrangement. However, they soon found that the arrangement had a downside. There was a dramatic increase in average finished goods inventory, with an accompanying decrease in inventory turns. For many companies, cycle times from the Chinese manufacturers grew to an average of more than 100 days from order to delivery. Customer metrics such as product availability slipped. Worse, the savings greatly diminished with rising fuel prices,

---

[3]T. L. Friedman, *The World Is Flat: A Brief History of the Twenty-First Century* (New York, NY: Farrar, Strauss, and Giroux, 2005).

growing production labor rates, and increasing inventory to curb the risk of disruption or order delay. In sum, many companies discovered a number of hidden costs and risks associated with offshoring.

As a result of the changing cost and risk structure, companies are beginning to consider locations alternative to Chinese/Asian sourcing. This seems to be leading full circle to a near-shoring trend worldwide. To stay in front of this changing global dynamic, supply chain professionals must understand the TCO for various supply chain alternatives to enable robust supply design solutions.

It Is "All About" the Shifting Total Cost of Ownership (TCO)

In general, when organizations pursue a cost-cutting strategy without giving much thought to enhancing customer service, they operate in a *cost world*, and not a *TCO* world. A short-term, cost-cutting mindset inevitably drives firms toward less-than-optimal decisions and strategies. Why did so many organizations find that the expected savings from offshoring and global sourcing initiatives did not materialize? An obvious answer is that these organizations simply did not do their homework. They focused on short-term cost reductions without considering all hidden costs associated with the TCO for offshoring decisions and the significant negative impact on supply chain responsiveness.

Our research shows that only a very small percentage of organizations consider some of the hard-to-compute costs that can severely hurt an organization's competitive advantage. These costs include the cost of lead time, the cost of flexibility, the cost of quality, the cost of lost sales, and of course the cost of added risk in a global environment. These costs give a whole new meaning to the phrase, "Distance matters." Organizations that pay attention to these costs will tend to focus more on near-shoring, or perhaps even reshoring, their supply chain operations. Near-shoring and reshoring are discussed more in the "Bending the Chain" section of this book. With this TCO framework as a backdrop, supply chain professionals can better understand the changing dynamics of the global supply chain environment.

Execution Is More Important Than Cost

While many organizations will acknowledge that they did not conduct a thorough TCO analysis prior to making global network decisions, they are reluctant to accept a more profound problem:

they cannot execute their global supply chain adequately in a global environment where competitive challenges rise exponentially. Going global increases cost, complexity, and risk. Managing these three aspects can be extremely challenging. Even if companies have estimated the total supply chain cost fairly well, they may be unable to handle supply chain complexity.

Worse yet, these companies may not realize or be prepared for the regional differences (cultural and otherwise) that exist in the location of the offshore activity. They may not fully understand the rules of this new global playing field or be prepared to manage the governmental rules and regulations of the country where they are planning to conduct such offshore activity. Even those organizations that have significant experience in offshoring may find that this is a never-ending and daunting process. The story of the second decade of the 21st century will be one of shifting cost and experience curves that are rendering the last 15 years of procurement, manufacturing, and supply chain location decisions obsolete.

Global Supply Chains Present Risk for Companies

Cost and execution challenges can keep supply chain professionals up at night, but what about risk in the global supply chain? Is near-shoring the solution for mitigating global risk? Part of the answer to this complex question depends on a complete TCO analysis. Additionally, companies must consider soft factors such as cultural or regulatory differences. To understand the impact of risk on global supply chains, Figure 2.1 defines important terms in the supply chain space. We have also included the following excerpt from Chapter 3:

> The supply chain arguably faces more risk than other areas of the company due to its global nature and systemic impact on the firm's financial performance. Risk is a fact of life for any supply chain, whether it is dealing with quality and safety challenges, supply shortages, legal issues, security problems, regulatory and environmental compliance, weather and natural disasters, or terrorism. There is always some element of risk.
>
> Companies with global supply chains face additional risks, including, but not limited to: longer lead times, supply disruptions due to global customs, foreign regulations and port

Some Terms

Supply Chain Network Design - (SCND) – The process for analysis and decision making on supply chain network investments. This includes key suppliers, manufacturing, warehouses, and technical centers. SCND includes capacity, SC capabilities, and "how/when" to use 3rd party partners.

Supply Chain(s) - (SC) – The end-to-end, integrated supply chain from the supplier's supplier to the consumer's shelf.

Supply Chain Organization – The holistic resources and teams required to deliver products to the consumer with excellence. This includes (but is not limited to) purchasing, manufacturing, engineering, process control, quality, safety/environmental, Innovation program management, ware - housing, transportation/distribution, and logistics (materials/production/ category/customer planning).

Figure 2.1 Benchmark companies have a broad, end to end definition of the SC

congestion, political and/or economic instability in a source country, and changes in economics such as exchange rates.

The scope and reach of the supply chain cries out for a formal, documented process to manage risk, but without a crisis to motivate action, risk planning often falls to the bottom of the priority list. The low priority for managing risk in companies is puzzling. After all, supply chain risk management is a very popular topic at conferences, and is written about extensively in books and articles. In spite of all of the discussion, we still see the vast majority of companies giving this topic much less attention than it deserves.

This risk apathy is driven by supply chain executives, who often find themselves at the center of the daily storm, striving to balance very demanding operational objectives while satisfying customers, cutting costs, and helping grow revenue. They must deliver results today, while working on capabilities that will make their companies competitive in the future. They operate in the same maelstrom of competing priorities and limited time as their executive peers—but their scope of activities is broader and they have less direct control over all the moving parts. In this environment, risk management receives much less priority than it should.

The repercussions of supply chain disruptions to the financial health of a company can be far-reaching and devastating. A study by Georgia Tech Professor Vinod Singhal and Kevin B. Hendricks emphasizes the negative consequences.[4] The study analyzed more than 800 supply chain disruptions that took place between 1989 and 2000. Firms that experienced major supply chain disruptions saw the following consequences over a 3-year period:

- 93% decrease in sales
- 33% to 40% lower shareholder returns
- 5% higher share price volatility
- 107% decline in operating income
- 114% return on assets decline

What Does All of This Mean for Global Supply Chains?

The preceding discussion is not meant to suggest that the globalizing supply chains, outsourcing, and offshoring are undesirable activities due to their riskiness. Rather, such activities must be considered from a more systematic, big-picture, TCO-based perspective. The objective is to create the optimal supply chain strategy based on business needs and total cost analysis.

In addition, our research supports the notion that global supply chains across the world will break into a series of demand and supply "pods" where regional procurement and manufacturing operations will supply the major demand centers of the area, at least for a significant percentage of production requirements. Clearly, some low-cost "commodity" items will continue to be procured from low-labor-cost regions across the globe. With a trend toward more regional activity, the question then becomes one of identifying the regional locations for offshoring or outsourcing. Should organizations return to procuring from, and manufacturing at, their domestic locations? If so, is the talent and infrastructure still there? Are the total costs and tax environment competitive? These questions are addressed in *Global Supply Chains: Evaluating Regions on an EPIC Framework*, and best

---

[4]K. Hendricks and V. Singhal, "An Empirical Analysis of the Effect of Supply Chain Disruptions on Long-Run Stock Price Performance and Equity Risk of the Firm," *Production and Operations Management* 14, no. 1 (2005): 35–52.

practices that companies are adapting for SCND (the process for determining the best supply chain solution for a business, including siting for key suppliers, manufacturing, warehousing, and technical centers) are discussed later in this chapter.

Multi-local is the term currently being used to describe this shift back toward domestic sourcing. Gartner, Inc., is a leading IT research and advisory company that publishes an annual "Supply Chain Top 25" list[5] identifying the leading organizations that excel in global supply chain management and highlighting their best practices. In a recent report, Gartner identified three major trends based on the practices of these top organizations:

1. Improved supply chain risk management and resilience
2. Supply chain simplification
3. A shift toward *multi-local* operations

The third trend, multi-local operations, relates to how these organizations are reassessing their sourcing and manufacturing network to then rebalance their supply network strategies. More specifically, "they are shifting from a centralized model, where these functions support global markets, to a regionalized approach, where capabilities are placed locally but architected globally." This multi-local trend supports the position on regional procurement and manufacturing.

The Gartner report identifies this trend toward multi-local locations as being driven by a number of factors: tax and government incentives, wage increases in some developing countries such as China, and an ever-increasing demand to be responsive to local markets. With respect to wages, the report notes that manufacturers are shifting capacity based on regional wage and logistics expense differentials even within emerging markets. Figure 2.2 provides a graphical illustration of the shift from Few/Global to Multiple/Local supply chains.

The Need for a Global Supply Chain Management Strategy

All of these complex issues cry out for a framework to inform global supply chain strategy. A policy of simply reacting to the dynamics of

---

[5]S. Aronow, "The Gartner Supply Chain Top 25 for 2014," May 21, 2014, http://www.gartner.com/technology/supply-chain/top25.jsp

Figure 2.2 Global sourcing decisions are shifting back to a balance of local and global design

the global environment will not cut it. Supply chains exist to serve customers. Therefore, the CEO and the executive team face the classic tradeoff of how to balance supply chain costs and inventory with customer service, a tradeoff that becomes incredibly difficult for a global supply chain. Supply chain managers face significant challenges walking this tight rope. Decisions affecting supply chain management are taken as part of corporate global strategy often without giving due attention to how such decisions can be implemented globally by the supply chain managers.

For many firms, we have found that top-level corporate strategies ignore the problem of managing global logistics or view it as an afterthought, a detail that can be accounted for eventually. However, supply chain issues are usually very significant; they can severely erode profit margins, return on invested capital, and shareholder value.

For some perspective, we typically see logistics-driven costs accounting for just under 10% of the gross domestic product (GDP), with overall global supply chain costs being a significantly higher percent. Assuming that this percentage is representative of the share of global logistics costs for an enterprise, a 5% error in estimating logistics costs in a $3 billion organization can result in a profit margin erosion of more than $275 million. That is not small change. If other supply chain costs such as order processing, materials acquisition and inventory, planning, financing, and information management are considered, the potential erosion in profit margin could be much higher.

Issues surrounding global supply chain strategies are often structural in nature, involving decisions on where to source material, locate a manufacturing facility, or open a retail center. For example, a decision to source material from a new location is often accompanied by eliminating material from an existing source. Such decisions cannot be reversed easily if the new offshore location fails to meet expectations. Because of this, some corporate strategies now place greater emphasis on the strategic importance of the global supply chain management process, but much additional progress is needed.

Building an efficient global supply chain, however, poses significant challenges. The manager has to juggle a multitude of conflicting objectives and contend with:

- Increasing consumer expectations on product quality and customer service
- Coordinating global supply chain partners to integrate supply and demand
- Global supply chain disruption risks
- Governmental rules and regulations in disparate countries
- Environmental concerns

While the first three concerns are present in every supply chain, the last two are probably more relevant to global supply chain. Although the United States lags behind Western Europe in environmental focus, environmental concerns are gaining momentum rapidly. In a survey by McKinsey on the challenges faced by supply chain managers,[6] environmental concerns were found to pose a fast-growing challenge. More than 21% of the respondents indicated that environmental concerns were their top challenge, nearly double the percentage from a survey conducted 3 years earlier.

To develop a global supply chain strategy, we recommend a logical and rigorous process that will take time and resources, but pay off many times over in the long run. In particular, we recommend the nine-step strategy development process found in the book *Supply Chain Transformation*.[7] This supply chain strategic

---

[6]"The Challenges Ahead for Supply Chains: McKinsey Global Survey Results," November 2010, http://www.mckinsey.com/insights/operations/the_challenges_ahead_for_supply_chains_mckinsey_global_survey_results

[7]P. Dittmann, *Supply Chain Transformation: Building and Executing an Integrated Supply Chain Strategy* (New York, NY: McGraw-Hill, 2012).

A GLOBAL SUPPLY CHAIN FRAMEWORK

Supply chain professionals who operate in the global environment need to be armed with a solid global supply chain management strategy. An important tool needed in developing this global supply chain strategy is the EPIC framework discussed in detail in the book *Global Supply Chains: Evaluating Regions on an Epic Framework.* We'll summarize the highlights of that framework in the section below.

Figure 2.3 A quality global SC strategy is essential

work requires a framework to assess the capabilities of supply chains throughout the world. This assessment is found in the EPIC research. The proper direction for global supply chain strategy is summarized in Figure 2.3.

EPIC Framework to Evaluate Global Regions

Global supply chain managers can benefit from a tool that helps them assess their supply chain strategies, identifying the strengths, weaknesses, opportunities, and threats of the different regions in the world. This tool is the EPIC framework, and it provides a structure for assessing various regions around the globe for supply chain readiness from an economic (E), political (P), infrastructural (I), and competence (C) perspective.

The EPIC framework defines and explains the environmental dimensions that impact the effectiveness of global supply chain management activities. It identifies the characteristics of these dimensions in each region of the world. The framework measures and assesses the level of maturity of a geographic region with respect to its ability to support supply chain activities. The four dimensions are, in turn, assessed using a set of variables associated with each dimension. Each one of these variables is assessed using a combination of quantitative and qualitative scores based on data drawn from a wide variety of data sources.

Assessing Supply Chain Readiness Across Regions of the Globe

Global supply chain managers can benefit from the EPIC framework because it helps them make better global supply chain location decisions in both emerging and mature markets. The EPIC framework provides a structure that allows managers to assess the readiness of global locations to support supply chain operations.

How the EPIC Framework Works

The EPIC framework assesses the key characteristics of a nation that are critical to managing efficient and effective supply chains. Each of these dimensions is evaluated using a number of variables to arrive at a weighted score for that dimension. In turn, the scores on these dimensions are used to arrive at a weighted score for the country. Each of the EPIC dimensions is described in the following.

Economy

The economy dimension assesses the economic output of the country, its potential for future growth, its ability to attract FDI, and how well it can generate a steady return on investments made in the country. The variables used to assess the economy dimension are the GDP and its growth rate, the population, FDI, exchange rate stability, consumer price inflation, and the balance of trade. These variables represent the potential opportunity that exists for organizations wishing to engage in supply chain activity in the country. For instance, the GDP of a country is largely determined by its industrial or service activity, which in turn significantly influences the level of supply chain activities.

Politics

The politics dimension assesses the political landscape with respect to how well it nurtures supply chain activity. The variables considered in the politics dimension include the ease of doing business, bureaucracy and corruption, the legal and regulatory framework, tariff barriers, the risk of political stability, and intellectual property rights. These variables influence the environment within which supply chains operate. For example, bureaucracy, corruption, stability of the political system, intellectual property rights, and hiring and firing laws significantly impact even the day-to-day operations in a supply chain. In many countries, it requires several weeks to receive customs clearance. Similarly, in certain regions transportation can be delayed or disrupted by "informal barriers" along the journey that require unforeseen payment before being allowed passage.

Politics is particularly important in the initial implementation phase of a supply chain project, as it requires managers to have knowledge of the country and ports from which products will be imported, the safest location for warehousing facilities, and so on. Costly delays can result from such issues as licensing, hiring, and environmental compliance. Furthermore, the encoding of cultural and historical norms in the laws of the nation form the legal framework for operations. Managers have consistently mentioned political issues as being among the most difficult when operating in a global setting.

Infrastructure

The infrastructure dimension tracks variables that strongly influence how supply chains in a country are managed and operated. It represents the potential for leveraging these activities. The variables considered in the infrastructure dimension can be broadly classified into three categories: physical, energy, and telecommunication infrastructures. The physical infrastructure covers the roadways, the railway network, and air and water transportation. The energy infrastructure is responsible for the supply of electricity and fuel. The telecommunications infrastructure is captured by the extent of telephonic and Internet-based activity.

Infrastructure has a direct impact on the economy as a whole and especially on supply chain performance. In Sub-Saharan Africa, for example, development efforts are recognizing that more than 50% of growth in the region is due to development of transport and telecommunications infrastructure. The tangible characteristics of a nation's transportation, utilities, and telecommunications infrastructure required to execute supply chain activities greatly affect supply chain performance. An effective ground transportation network greatly facilitates cost-effective movement of product among sourcing, manufacturing, and market areas. Air and seaport facilities are essential to support global trade by efficiently and effectively moving materials into and out of the region. Investment in infrastructure is an element that can be tracked and is a strong predictor of business growth in a nation or region.

Decisions on infrastructure development also require a sound understanding of geography. Roads over tall mountains, across large deserts, and through jungles and marshes generally are not very effective or efficient. In addition, access to stable electricity, water, and telecommunications is essential. For example, many supply chain managers in emerging economies spend significant time and/ or money arranging for power generation. Even if a firm does not operate in an emerging market, one or more of its suppliers is likely to operate there. As a result, supply chain managers must be knowledgeable about the conditions in which those suppliers operate so as to ensure top overall supply chain performance.

Competence

The competence dimension assesses the general supply chain skill levels of both the work force and the logistics industry within and out from a country that is a potential part of an organization's supply chain. Variables include:

- Labor productivity
- Labor relations
- Availability of skilled labor
- Education level of line staff and management
- Availability and competence of the existing logistics service industry
- Speed with which customs and security clearances take place

Competence is another dimension with huge direct impact on supply chain performance. Availability of labor, labor productivity, and the sophistication of supply chain support available through the logistics industry in a nation affect the ability to run high-performing supply chains. Mastery of the tangible requirements for supply chain operations is a necessary but not sufficient condition for success.

Supply chain managers must also explore the conditions related to so-called soft issues that culture, history, population, and politics have on supply chain operations. Leading and managing a local workforce is a key success factor in designing and executing supply chain solutions. How people in a region regard work makes a difference. People do not have the same skills, the same references, the same education, or the same hopes from one region to another in the same country, and certainly not across nations or regions. Such issues impact labor force management, attendance, attrition, skill levels, and much more. Performance objectives are not the same by region: in one region level of service may be the requirement for success, whereas in another efficient management of inventory may be key.

The Regional Assessments

Regional assessments excerpted from the EPIC book are shown in Table 2.2. The table provides the EPIC assessment for 55 countries included in the analysis. It is organized according to the countries or regions with the highest-rated overall EPIC assessment.

Table 2.2 The Key Variables Important to These Assessments

Country	Economy	Politics	Infrastructure	Competence	Overall
Hong Kong	4.00	5.00	5.00	5.00	5.00
Singapore	4.00	4.50	5.00	5.00	4.50
Germany	4.00	4.50	5.00	4.50	4.50
Netherlands	3.50	4.50	5.00	4.50	4.50
Sweden	3.50	5.00	5.00	4.50	4.50
United Kingdom	3.50	5.00	5.00	5.00	4.50
United States of America	4.00	4.50	4.00	5.00	4.50
Canada	4.00	5.00	4.00	4.50	4.50
Japan	3.50	4.50	4.00	4.50	4.00
South Korea	4.00	4.00	4.50	3.50	4.00
Taiwan	4.00	4.00	4.00	4.50	4.00
Malaysia	4.00	3.00	4.00	4.50	4.00
Australia	4.00	4.50	4.00	4.00	4.00
Saudi Arabia	4.00	3.00	4.00	4.00	4.00
United Arab Emirates	3.50	4.00	5.00	5.00	4.00
Austria	3.50	5.00	4.50	4.50	4.00
France	4.00	4.00	4.50	3.50	4.00
Chile	4.00	4.00	3.50	3.50	4.00
China	5.00	2.50	2.50	4.00	3.50
Thailand	4.50	2.50	1.00	3.50	3.50
Israel	3.50	4.00	3.50	4.00	3.50

Table 2.2 (*Continued*)

Country	Economy	Politics	Infrastructure	Competence	Overall
Czech Republic	3.50	4.00	3.50	3.00	3.50
India	4.00	2.50	1.50	3.50	3.00
South Africa	3.50	3.00	3.00	3.00	3.00
Italy	3.50	3.50	3.00	2.50	3.00
Spain	4.00	3.50	4.00	3.00	3.00
Turkey	4.00	2.50	3.00	3.00	3.00
Hungary	2.50	3.00	3.00	2.50	3.00
Poland	4.00	4.00	2.00	3.00	3.00
Mexico	4.00	2.50	2.50	3.00	3.00
Panama	3.50	2.50	4.00	2.50	3.00
Brazil	4.50	2.50	1.50	3.00	3.00
Peru	4.00	2.00	2.00	2.50	3.00
Uruguay	3.00	3.50	3.00	1.50	3.00
Indonesia	4.00	1.50	1.50	3.00	2.50
Philippines	4.00	1.00	1.00	3.00	2.50
Vietnam	3.50	2.50	4.00	2.50	2.50
Egypt	4.00	2.00	2.00	1.50	2.50
Russia	4.00	1.50	1.50	1.50	2.50
Costa Rica	2.50	3.50	1.50	3.00	2.50
Colombia	4.00	2.00	1.50	2.50	2.50
Pakistan	3.00	0.50	1.50	2.50	2.00
Romania	3.00	2.50	0.50	2.00	2.00
Ukraine	3.50	0.50	1.50	2.00	2.00
Argentina	3.50	1.00	1.50	2.00	2.00
Bangladesh	3.00	0.50	0.25	1.50	1.50
Algeria	3.00	0.50	1.50	0.25	1.50
Ethiopia	3.00	1.50	1.00	0.50	1.50
Kenya	2.50	1.00	1.50	1.50	1.50
Nigeria	4.00	0.50	0.50	1.00	1.50
Myanmar	3.00	0.25	0.25	0.50	1.00
Angola	3.00	0.25	0.25	0.50	1.00
Venezuela	2.50	0.25	0.50	0.50	1.00
Dem. Rep. of the Congo	3.00	0.00	0.00	0.00	0.50
Sudan	2.50	0.00	0.50	0.25	0.50

EPIC Framework and the Supply Chain Network Design (SCND) Model

These EPIC assessments can be very helpful in completing a first pass at a global strategic design. Now, let us take it down a level with the introduction of the SCND model. In Table 2.3, we introduce a list of practical and more detailed issues that are aligned with the variables in the EPIC framework, and must be considered in a global SCND. Figure 2.4, meanwhile, concludes this section by transitioning to the summary of the EPIC Assessment's key themes.

Summary of Key Themes From the EPIC Assessment

- Supply chains have undergone a series of phases and transitions over the ages, from trading supply chains to manufacturing supply chains to the current era of global, IT-enabled supply chains. These transitions have been fueled by political and technological innovations that include deregulation of investment and trade laws, interchangeability of components, improved methods of transportation, mechanization, telecommunication, and the Internet.

- These political and technological innovations have resulted in global economic power moving from Asia to Europe beginning around the middle of the 18th century, and on to North America following World War II. However, recent trends suggest that the balance of economic power is either moving back to Asia or at least being leveled across the Americas, Europe, and Asia. Supply chain professionals must have a robust *TCO* process to stay ahead of the changes in the dynamic global environment.

- The world is becoming flatter as barriers to free-market capitalism are removed, the use of the Internet becomes more widespread, and there is an increased flow of goods across borders. However, the extent to which the playing field is being leveled is open to debate. Questions remain about whether the flattening trend could reverse course. Stranger things have happened!

As global supply chains proliferate, organizations in one country will increasingly depend on organizations from other countries to

Table 2.3 Supply Chain Network Design Issues, Key Variables, and the EPIC Dimensions They Fall Under.

Supply Chain Network Design Issues/SCND	Key Variables	Dimension(s)
Decisions on product design	Intellectual Property Rights	Politics
Design school and champion	Education Level	Competence
E-commerce vs. retail store	Population Size Intellectual Property Rights Education Level Logistics Competence	Economy Politics Competence
Logistics network design	Transportation Infrastructure	Infrastructure
Manufacturing location	Foreign Direct Investment Exchange Rate Stability & CPI	Economy
R&D center	Intellectual Property Rights Education Level	Politics Competence
Retail store location	GDP and GDP Growth Rate Population Size Ease of Doing Business Legal & Regulatory Framework Risk of Political Stability Telecommunication & Connectivity Education Level	Economy Politics Infrastructure Competence
Sales channel—direct sales stores vs. distributors	Population Size	Economy
Sourcing & manufacturing location	Balance of Trade Ease of Doing Business Legal & Regulatory Framework Risk of Political Stability	Economy Politics
Sourcing, manufacturing, and logistics location	Utility Infrastructure (Electricity) Telecommunication & Connectivity Labor Relations Education Level Logistics Competence Customs & Security	Infrastructure Competence
Supply network—node location	GDP and GDP Growth Rate Ease of Doing Business Legal & Regulatory Framework Risk of Political Stability	Economy Politics

> The SCND issues form the basis of the best practice discussion later in this paper. From this discussion we'll see how leading Fortune 500 companies address these issues. Before getting into the best practices, let's first summarize the key themes from research.

Figure 2.4 Benchmark supply chains follow these best practices for SC network decision making

either supply material or market their products. Organizations should therefore position themselves to take advantage of such codependencies to further their competitive position in the marketplace. They must question on which regions of the world they should focus their attentions when developing such collaboration.

Our research supports the notion that global supply chains across the world will break into a series of demand and supply pods where regional procurement and manufacturing operations will supply the major demand centers of the area, at least for a significant percentage of production requirements.

In addition, three predominant themes emerge from the EPIC assessment that relate to key supply chain decision areas covering demand markets, sourcing and manufacturing, and global trade and logistics.

1. *Demand market trends*: Despite the fact that other regions of the world are closing the gap in economic activity, strong consumer markets for finished goods remain in the United States and Canada, the European Union (although at a lower level than pre-2009 heights), Japan, South Korea, and Taiwan, as well as in the large emerging markets of the BRIC nations (Brazil, Russia, India, and China). In these countries, product volumes, diversity, and short product life cycles are key for consumer goods managers, in particular for managers of fast-moving consumer goods. Other nations with rising consumer markets include Mexico, Turkey, Saudi Arabia, Colombia, South Africa, Indonesia, Malaysia, and Thailand. In all of these countries, the speed of development, changing volume requirements, and varying geographies and transportation infrastructures are the most challenging issues.

2. *Sourcing and manufacturing*: Significant reengineering of supply chain networks is currently underway. The top emerging areas of opportunity for sourcing, manufacturing, and logistics to support regional and/or global consumer markets include Vietnam, Malaysia, India, Chile, Colombia, Uruguay, Brazil, Mexico, Costa Rica, Poland, Czech Republic, Slovakia, Nigeria, South Africa, Kenya, Germany, and the southern and western regions of the United States. In addition, many opportunities are just beginning to emerge in Africa, largely supported by infrastructure investment from China.[8] Even in certain regions of Africa, where the manufacturing network may not meet requirements for worldwide distribution, factories are capable of efficiently and effectively serving as a source of supply for products delivered within the region.[9]

3. *Global trade and logistics hubs*: The changes in market, sourcing, and manufacturing locations will predicate changing trade lanes among business nodes. Usual locations for establishing global trade and logistics hubs, as well as minor assembly, packaging, and/or redistribution facilities include Hong Kong, Singapore, Shanghai, and Rotterdam. In addition, similar opportunities are emerging in locations such as the UAE (specifically Dubai), Panama, Colombia, Saudi Arabia (thanks to the developing trans-Arabian highway), South Africa, Egypt (assuming political stability returns in the near future), Algeria, and Morocco. Even within regions, trade flows are shifting to highlight new areas of focus for assembly and logistics operations. In Europe, the center of gravity for trade flows is slowly shifting from a Western-oriented logistics network to one that is more centrally focused on the continent. In North America, flows are slowly shifting from an east–west or west–east axis to a more south–north access as Mexican ports and manufacturing centers gain in prevalence. The anticipated opening of the new Panama Canal in 2016 also promises to increase trade volumes in Gulf of Mexico and Southeastern US ports, further strengthening that trend.

---

[8]T. Paulais, *Financing Africa's Cities: The Imperative of Local Investment* (Washington, DC: French Agency for Development and the Work Bank, 2012).

[9]V. Foster and C. Briceño-Garmendia, *African Infrastructure: A Time for Transformation* (Washington, DC: French Agency for Development and The Work Bank, 2010).

Best Practices

Equipped with the EPIC framework, we talked with 10 leading supply chain companies that have a best-in-class capability not only in managing large, complex global supply chains, but also in global SCND. These companies came from CPG, chemical, luxury, materials refining, food, supply chain services, and health care industries.

In the interviews conducted to identify best practices, the best-in-class supply chain organizations addressed two important questions:

1. How should leaders make the network location decisions for warehouses, key suppliers, manufacturing, and technical centers (also known as the SCND process)?
2. As a leader and owner of a complex, global supply chain, how do you best manage it?

Based on these questions, we developed the following two sets of best practices:

1. SCND
2. Managing complex, global supply chains

SCND—Global Sourcing Analysis and Decision Making

In making global sourcing decisions, our best-in-class companies followed five best practices:

1. Supply chain decisions are *strategy driven*
2. These firms strive for *scale*
3. Supply chain organizations have a core competence in *fact-based, quantitative analysis*
4. Supply chain decisions are *net present value* (NPV) and *TCO*-based
5. Supply chain decisions are based on *holistic product designs* appropriate for the global environment

We discuss each of these five best practices below, and give practical examples for each.

1. Strategy driven

Supply chain sourcing and network decisions need to support the strategy of your business. Clearly, supply chain leaders want to design supply networks that provide a competitive advantage, based on the strategic needs of the business and the competing market/products.

Unfortunately, this is a major challenge for supply chain leaders. The issue is not a lack of desire, but a lack of resources available to supply chain professionals to understand the global market trends. In fact, one general manager states, "I cannot predict the future, especially in the global environment—you build what you think you need now."

Obviously, this is dangerous. As a first step, supply chain leaders must work with business leadership, marketing, and R&D to document a clear picture of the long-term business plan (goals, strategies, priorities, regional splits, sources of volume, categories, and products). Frequently, the supply chain leader must facilitate this process. *This is step 1 and the cornerstone of SCND.*

Beyond the long-range business plan, the overall business leadership team needs to provide clear direction in several key areas that will drive the global supply chain strategy. These include:

- *Supply chain risk.* What are your key supply chain risks (natural disaster, supplier/plant failures, governmental/regulatory, security/terrorism, financial/currency, strike/labor, cyber, etc.)? Is the company prepared to provide capital, resources, and expenses to mitigate the critical risks? This is a key SCND input. It provides the basis to address questions about single sourcing, supplier partnerships, capital requirements, and countries to avoid. The EPIC framework provides the necessary supporting, region/country insight for this analysis.

- *Responsiveness.* Most general managers start with an expectation for 100% customer service. However, supply chain professionals need to guide the business regarding the fundamental SCND tradeoffs involved. If your business requirements dictate very few, global key suppliers and manufacturing plants, you will have a longer supply chain cycle time. This means your inventory will be higher, your responsiveness to ongoing business variation/demand will be lower, and your responsiveness to launching new product initiatives simultaneously around the work could be reduced. It is best when these tradeoffs are clearly understood and aligned multifunctionally.

- *New product innovation.* The investment in the supply chain network creates by its very nature parameters for R&D to develop new products. To launch a new product within limited capital and expense budgets, R&D will need to largely utilize existing equipment, processes, suppliers, and materials. To launch "new to the world" disruptive product innovation (requiring new supply chains), the incremental revenue (market share) needs to be sufficient to create a strong return on the investment. Therefore, the supply chain leader and the R&D leader need to be closely aligned. The technical community (supply chain and R&D leaders) should go to the business leaders with one voice on supply chain capability, product design parameters, and the capacity needed for multi-year new product initiative pipeline.

Examples:

- A large, global company producing beauty care products has a manufacturing design for a plant in North Carolina and another in South Africa. The global revenue is primarily in North America (NA), Western Europe (WE), and South Africa. This SCND was based upon the South African plant supplying WE and South Africa with global products. Unfortunately, the demand in the South African market for these global products was insufficient to meet local business goals. The local South African GM had local profit responsibility. Therefore, the South African plant utilized the majority of its capacity for local products. That required North Carolina to source the product for WE. Due to its high-cost structure, this negatively impacted WE business. This global plant network design clearly was not strategic or business driven. This issue is currently being solved through acquisition of WE production facilities.

- A large, global CPG company making detergent products struggled with predicting product design and product forms in various parts of the world. The lack of clarity on future product forms and the long lead time to react created significant losses resulting from investments in powder laundry detergents, then investments in liquid laundry detergents, and most recently investments in laundry "pods." Clearly, the supply chain failed to provide acceptable service to customers in this environment. This is a strong example of why general managers must define the future and synchronize capacity investments and global supply chain capabilities with new product innovation.

- A global chemicals company requires its procurement managers to submit at least two business improvement proposals per year (in the corporate/marketing format). This expectation is included formally in the manager's work plan and bonus assessment. The supply chain leader wants a strong culture of strategic/business-focused supply chain procurement managers.

- A large, global pharmaceutical company struggled with manufacturing investments. Based on business leadership direction, the supply chain built two manufacturing plants (well over $100 million in capital) to produce a new drug that ultimately was not approved for consumer safety and a new drug that was non-competitive in the market due to consumers' issues with application. These plant investment decisions cost the company tens of millions of dollars.

- A large, global chemical company is now requiring that all multifunctional business teams (making profit/loss decisions) have two important members from the supply chain organization: purchasing and logistics. This ensures that the business decisions incorporate all the key business components (including supply chain implications).

2. Scale

Arguably, the single biggest benefit of achieving a quality SCND is the increased understanding that leaders gain about the supply chains they manage. When that happens, major benefits in global scale can occur. In fact, the "gem" of SCND is when you find scale that can be leveraged for lower cost/cash, higher quality, shorter supply chain time, and improved customer service (a win across the board). This scale could be within your global supply chain, within your company (across other categories), within global industry (e.g., partnerships with suppliers, 3PLs, competitors), or in other industries having similar equipment, processes, systems, and/or technology.

Examples:

- A large, global consumer products company had significant growth in Central/Eastern Europe. Unfortunately, two of the business units in this large company failed to work together. One business unit built a new, state-of-the-art plant in Romania and the other independently built a new, equally impressive plant in Poland. This is a classic example of not utilizing the scale within the company. Both sites have had successful "start-ups" and early success, but both suffer from lost cost opportunities by having excess capacity. A corporate decision for one plant in Central/Eastern Europe would have saved precious supply chain resources.

- A large, global chemical company is driving scale through its acquisition process. Historically, the company would integrate the new acquisitions IT, management, key suppliers, and cultural systems based on the perceived value of the project (priority setting). This led to a series of acquisitions not meeting published financial goals. Today, as a part of the acquisition financial and implementation plan, the company immediately transitions the operations to its SAP IT systems, retains managers with technical mastery (separating other managers that cannot quickly recreate the company culture), and transitions applicable suppliers to the current supply base. This requires more scale work early in the transition but has led to consistently meeting the acquisition financial goals.

- Finally, a large, global CPG company designs its Asia manufacturing site for multiple category products. In "Western countries," the plants are designed by category to provide organizational focus to deliver the business goals. In Asia, the major regional challenges include talent retention (including recruiting/development) and sufficient scale to impact governmental/regulatory support. The larger Asia sites help address these challenges.

3. Analysis Mastery

Many of the examples included in this best practice section highlight failures in SCND that resulted from incomplete assessments of TCO, lack of scale, and product design. Inadequate SCND was caused by the absence of people and resources to complete a high-quality analysis. We have found that many companies initially believe that if you team a smart, talented manager with strong financial support and open access to the supply chain senior executive, you will complete high-quality SCND analysis. These companies have learned the hard way that the global supply chain landscape is extremely complex. The EPIC framework provides insight into the complexity of this analysis. The world is changing so fast that the EPIC analysis is a continuous process.

We found that the best-in-class supply chains:

- Have a dedicated business/SCND analytics department, SCND masters that maintain analytical technology, or external partners to lead the analysis. They have data systems and analytical tools designed for accuracy and useful information. The masters have strong connections with access to the latest and most accurate tax structures, duties, customs, transportation costs, and more.

- Start the SCND work with a detailed mapping of the "end-to-end" supply chain (current and proposed). This step provides two important outcomes. First, it provides a visual training for leadership and the analysis team on global supply chain. Second, in our new world of complex, global supply chains, detailed mapping ensures that the right questions are addressed in the analysis.

- Spend at least 15% of the analysis time on helping leadership define the problem, which is rarely clear in the initial request. They spend approximately 35% of the time completing the analysis (using the tools and systems that masters can maintain) and roughly 50% of the time on sensitivity analysis. This breakdown is visualized in Figure 2.5.

Sensitivity analysis is critical. It includes:

- "What if" analysis (impact of key risks including competitive reaction, etc.)

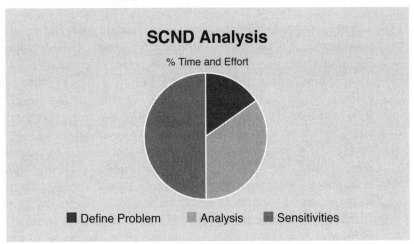

Figure 2.5 SC decision analysis focused on the decision risks and sensitivities

- NPV range diagrams (some companies call these "tornado diagrams") that show the NPV range for different volumes, capital spending, energy prices, etc.

4. TCO/NPV Based

In the introduction, the need for a TCO analysis and an NPV analysis on all key decisions is discussed extensively. This is clearly an area that has broad alignment and organization support. All the benchmark companies we studied use NPV as the primary measure to compare options in the SCND analysis. This ensures that the best overall value and long-term decisions are being made. None of the benchmark companies were having issues with this element of the analysis.

In addition, all the benchmark companies are using a variation of TCO in the SCND analysis. The challenge is in the execution of TCO analysis. Assessing all elements of cost for the complex global supply chain has been difficult. Table 2.4 defines categories of costs and potential examples of each. It also has been extremely difficult to determine the true cost of key supply metrics including time, responsiveness, supply chain risk, quality, and inventory. The recent "reshoring" trend and a more balanced strategy/business-driven network design largely corrected many incomplete, initial SCND and TCO analyses. Unfortunately, most of these companies have examples of poor decisions when the total cost analysis was incomplete.

Examples:

- A large, global CPG company completed an analysis to shift production of a body wash product to Mexico. This decision was made at a time when there was heavy discussion in the company regarding leveraging NAFTA and utilizing countries with significant wage rate advantages. The analysis was largely based on manufacturing cost alone. After a decade, the company determined that this initial analysis was flawed. Body wash (large, heavy bottles) has a large transportation cost/unit and a large overall supply chain cost as percentage of total revenue. It requires a responsive supply chain to handle the significant level of customer promotions, and it is sensitive to a high level of inventory. (Note: These types of supply chain/product attributes and how they impact SCND are discussed below in the chapter summary.) When the company completed a second

Table 2.4 Key Items to Include in TCO Analysis

Examples of costs that should go into TCO calculations	
Transportation Costs	**Cost of Schedule Non-Compliance**
• Freight Costs	• Expedite Costs (Air Freight)
• Custom Duties and Tariffs	• Stockout and Lost Sales Cost
• Brokerage Fees	
Cost of Additional Inventory	**Payment Terms**
• Pipeline Inventory	• Cash Discounts
• Safety Stock	• Payment Terms in Days Outstanding
Cost of Quality and Obsolescence	**Cost of Administration**
• Factor for Warranty Claims	• Offshore Supplier Qualification Costs
• Scrap and Obsolescence Risk	• Cost of Administration Trips to Offshore Location
• Inspection and Disposal Costs per Shipment	• Cost of Communication
Risk Costs	**Cost of Responsiveness**
• Currency Risk	• Lead Time Costs
• Country Risk	• Flexibility Costs
• Competition (Intellectual Property) Risk	• Longer Quality Feedback Loop Costs
• Job Switching Risk	• Port Congestion Costs
• Transportation Disruption Risk	

SCND, it appropriately moved the body wash business back to a US plant. This enormous cost could have been avoided if a proper TCO analysis had been done in the beginning.

- A high-end, global, "luxury product" company utilized a few global manufacturing and warehousing sites (see this section's summary for SCND on product with a very low supply chain cost as percentage of revenue and low/manual technology). The supply chain focused on locating the few, global sites in low-wage-rate countries in Asia to further improve margins. Initially, the manufacturing sites were located in China. As the company learned more about the increasing Chinese wage rate trends and understood Chinese custom/duty cost, the sites were moved to more cost effective locations in South East Asia, such as Vietnam.

- A large household product company completed an SCND on a new product that included a device, a fabric, and a liquid cleaner.

The cost analysis focused on each component in isolation. The accepted practice was to source simple devices from Asia (low-wage markets), which was done. They sourced the fabric component from a single global manufacturing site in Germany and the liquid product from the United States. The final SCND included a multination global supply network (China, Germany, and the United States). This proved to be a significant business problem as the "lead" market chosen was the United States. This product required quick response time (customer and new product driven) and low levels of inventory (new product innovation). The supply chain was long and slow and had to be redesigned.

5. Holistic Product Design

The benchmark companies we interviewed have frequently learned and developed their global product designs the hard way: mistakes. In the supply chain introduction, we discussed the learning associated with inaccurate total cost analysis, or not appropriately valuing supply chain time and responsiveness.

A second opportunity can be found in holistic product design. Businesses under extreme Wall Street pressure to grow profit and revenue now have made relatively quick decisions to expand current products globally. These decisions create massive supply chain projects, significant resources, and large capital investments. Supply chain leaders must drive the business and R&D leaders to do the homework on whether consumers will embrace these global products and deliver the early forecasts/global expansion goals.

The major challenges have been entering developing countries where consumers have significantly less disposable income. Typically, lower cost, good-quality variations of the global (Western world–based designs) are required to win with these consumers. This enables the supply chain to use the same global platform (standard, materials, equipment, processes) but in a way that delivers products affordable to their new consumers. This same principle of "overdesign" applies broadly to all new products. R&D and Marketing must determine what product attributes the consumer is willing to pay for and which ones they will not (avoiding expensive supply chain design rework).

Examples:

- A global consumer products company launched a new product. The marketing department was convinced that the carton color was vital for the new product's success. The chosen shade of red was only available from an African supplier (Ghana) at a premium cost that required additional inventory investments and longer supply chain cycle time. After the product was launched, the business determined that the new product's critical consumer attributes were price, product performance, and availability on shelf. The unique African dye cost the company millions of dollars in wasted global supply chain cost that could have been spent on the three consumer-driven needs or utilized for profit.

- A global, personal care products company had technical mastery of filling shampoo, hair conditioner, and skin care products in plastic bottles. When they chose to enter the massive India market, they launched with the products they knew best. Quickly the company realized that the India market has two components: one market that is largely similar to Western habits (large cities, large retailers) and a much bigger market with different consumer habits in the rural areas of India. They quickly developed a second package that fit into the rural outlets (small carts, small product "huts"). The company decided to reapply its individual sachets packets (designed for sampling in Western markets). The India rural consumer utilized these single-use product forms to match their needs. After launching the "sampling" sachets, the company determined they could not compete on price with the local competitors. The company was forced to study the market, competitors, and consumers to design low-cost sachets that met actual needs. This in-market, iterative process to define consumer needs created significant rework of the company's Indian supply chain.

Best Practices for Managing a Complex, Global Supply Chain

In our discussions with the best-in-class supply chains, we also saw five best practices in managing their complex global supply chains:

1. An effective *global S&OP* process
2. A *process to manage complexity*, especially product complexity
3. Strong *supplier collaborative partnerships*
4. A *talented team* on the ground
5. *Clear visibility* throughout their global supply chain and *rapid response capability*

We discuss each of these five best practice areas below and give practical examples for each.

1. Global Sales & Operations Planning

Sales & Operations Planning (S&OP, or integrated business planning) has been discussed in supply chain journals for decades. Many companies have an S&OP system and are working on improving their system to create a single-number, integrated business plan for 100% of the organization to utilize (marketing, sales, finance, and the supply chain) each month. Global S&OP is included in this section primarily due to the difficult nature of a truly global decision.

The same factors are driving exponential growth in supply chain complexity and making our partners' (marketing, sales, finance, R&D) work more complex. Frequently, profit centers in global companies are managed by country or region. This is driven primarily by:

- Better short-term results with profit managers close to their consumers
- Government laws, culture, and local financial practices that make local ownership more efficient

This creates a challenge for global supply chain leaders, however. As they work to manage the new world environment, they develop strategies for global suppliers, global manufacturing and warehousing, and sharing of assets among regions. The need for global decision making (S&OP) increases. We need leaders who can make decisions based on allocation of limited finished product and materials. Should the product go to the highest margin regions? Should it go to regions with the highest current marketing spending? Or should it flow to regions launching new products?

This issue is compounded by the practice of having the local corporate officer own the local supply chain assets. Is it a conflict of interest for local general managers to decide that their country is the first priority during a shortage, or that investments in local plants will be financially allocated to all other regions when the product is exported? Best-in-class companies clearly define accountability for critical decisions. For each business category, there is visible accountability for the global decision authority.

Examples:

- A large CPG company launched specially designed products to improve the application of its liquid/cream products. These unique products were single-sourced from Asia since they were low cost and very sensitive to wage rates. The North American market was the highest volume, highest profit region globally. By default, the Chinese general manager was the global decision maker. He owned the profit and the supply chain assets were in his region. During a period of tight supply, he decided to allocate available product to the Chinese market. The global supply chain leaders had to work this issue with North America, China, and Global Leadership. The process was inefficient and time consuming. It required multiple layers of the company leadership to find a person who could make a global, "shareholder"-based decision. This created much churn, confusion, and wasted resources.

- A large, global beauty products company was completing a major aerosol manufacturing sourcing change from the United States to Africa. This was a global decision. The local leadership in Africa teamed with R&D to decide on significant new product changes (formula, package) during the transition. The logic was that it would be easier for R&D and less costly for the local market. This proved to be a major business disruption as the project initially failed, crippled by its huge complexity. It created massive customer service outages (lost revenue). The supply chain leaders knew this was too high a risk but were unable to find a global decision maker willing to assess the total, corporate risk. The best approach would be for the business team to consider all of the data from every function. If the company had done this, the business damage from massive customer service defects would have been avoided. The supply chain leader must "step up" and demand that the decision processes are multiple function to ensure the best decisions.

- A large, global food company manages complex, global supply chains involving development of crop investments. Two of their high profit spices (vanilla and cinnamon) have significant looming capacity shortages. The crop planting investment requires a 15-year lead time (harvest when plants grow to maturity). This 15-year forecast horizon has changed the companies S&OP processes and has required new data systems.

- A global education company supplies products to enable learning. In the last decade, these products frequently use the latest IT technology. This is a new area for the company. Product obsolescence (due to the frequency of technology changes) is now a huge factor in profitability. Therefore, the business's global S&OP process has been renewed to ensure stronger alignment on "single-number" supply plans.

2. Process to Manage Complexity

We found that leading companies are forming or already have programs to handle the exponentially increasing levels of supply chain complexity driven by longer, global supply chains, increasing government/regulatory laws/guidelines, acquisitions, increasing levels of new product initiatives, and complex product designs (high level of stock keeping units or SKUs). Most of the best processes to manage complexity center around four strategic action plans:

- Standardization of equipment, processes, services, products, and systems
- Simplification of suppliers, materials, specifications
- Creation of scale by leveraging the company's scale, 3PL provider scale (industry), supplier scale (partnerships), or corporate joint ventures
- Implementation of an IT-enabled solution (the days of complex, global supply chains solving the current challenges without high-quality, real-time, integrated information systems has been over for some time)

Platform management is the methodology led by the technical community (supply chain and R&D) to systemically manage much of this global complexity. It was developed in high-capital industries (auto, airline, and heavy equipment) and focuses on driving out non–value added cost by leveraging robust process control, standardization, simplification, and early multifunctional technical community involvement in investments.

Examples:

- A large, global food company is growing through acquisition. This is exponentially driving complexity. In addition, the firm faces high global transportation costs and the absolute requirement to maintain product freshness. The appropriate SCND would invest in many local plants, key suppliers, and warehouses, but that creates enormous supply chain complexity. For example, the company has more than 4,000 global suppliers, over 700 product specifications for 1 product category in Europe alone, and more than 140 plants (over 30 of those in Europe). To manage that amount of complexity, the company and its supply chain leaders have deployed an aggressive scale strategy. This strategy involves creating scale by stretching equipment/product/services standardization goals, setting material/supplier harmonization goals, and focusing on manufacturing/warehouse productivity by rationalizing and improving the capacity of assets. In effect, the firm used platform management concepts and tools to drive the scale. This created a steep change in cost/cash, quality, new product innovation speed, and customer service by creating scale enhanced by the elimination of complexity that consumers did not value.

- A large, global pharmaceuticals company renewed its product life cycle management (PLCM) system. Several key prescription drugs moved out of patent protection. This transition created significant challenges on managing the "end-of-life" process. Historically, this event has caused major gaps in the revenue/profit progress within the company. The pharmaceutical company is learning from the high-capital/high-technology business that created PLCM methodology. The "end-of-life" process focuses on how to use these products in the most productive manner and how to make the right pretransition planning to optimize profits if the product cannot be reused.

3. Supplier Partnerships (Materials, Equipment)

A proven approach to supply chain complexity is the creation of supplier partnerships with strategic suppliers. The benchmark supply chains are expanding these partnerships as supply chains become even more complex. We have found that the focus is on the following areas:

- Creating partnerships with a larger set of strategic suppliers (increasing the number of partners)
- Expanding the partners beyond strategic material suppliers to equipment suppliers, 3PLs, service providers, etc.

- Starting to expand the concept to industry partners (driving out broader levels of waste and inefficiency)
- Expanding the focus to communities and sustainability

Examples:

- A large, global food company is creating supplier partnerships with local farmers and growers. The partnerships include company investment in local schools and community facilities. This is a broader definition of partnership but creates value through stronger teamwork and trust.
- A large, global company supplying personal care products is using its strategic material supplier partners to start developing new chemicals and packaging materials. This has caused significant cultural changes in the company (especially R&D), but has enabled the suppliers with the technical mastery to actively participate in new product improvements.
- A large, global food company revitalized material supplier partnerships with their equipment vendors. As the supply chain has implemented platform management, the manufacturing equipment/processes are globally standard. This enables significant capital reductions through partnerships with equipment suppliers. Blanket equipment orders can now be developed, allowing the equipment suppliers to level the workload and drive productivity and cost improvement.
- The global supply chain organization in one leading company is initiating discussions on an industry-wide basis. Can the company partner with its competitors to form industry standard secondary packaging? The idea is that secondary packaging is a "non–value added" cost from a consumer point of view. The most efficient solution is an industry standard for all similar products to eliminate warehouse, transportation, and customer logistics costs due to variation in design. This is a stretching goal and has significant challenges to its implementation, but it is a great example of how benchmark supply chain organizations are looking to create a new level of supply chain partners.

4. Talented, "On-the-Ground" Team

In the beginning, benchmark supply chains viewed global expansion as a test. Therefore, supply chain activities in new regions were frequently managed with corporate resources and contractors in those regions. This approach caused significant problems (slow responsiveness, high inventory, and limited capability with local regulatory agencies). Today, the supply chain organizations in our benchmark companies use the following principles:

- Hire a local "on-the-ground" team to manage the business
- Establish local budgets with a higher percent of training and development funding to support the new organization
- Staff with a sufficient level of "ex-patriot" (expat) managers to provide support for the new team
- Institute a high level of coaching for new employees
- Establish a high standard for corporate cultural expectations ("one global culture")
- Create a flexible pay system based on local practices to ensure retention of valued employees
- Assign additional corporate resources/coaches to ensure the success of the new "on-the-ground" teams
- Co-locate with local marketing and sales offices when possible to drive teamwork on solving local business challenges

Examples:

- A global materials (light weight metals) supplier struggled with quality variation, transportation delays, product accounting, and confidentiality agreements in China. Managing this from a Western country proved ineffective. The company invested in an "on-the-ground" Chinese team. This has paid strong dividends as the quality, cost, and competitive results have significantly improved. The local resources were more effective at resolving issues based on their local understanding of supplier capabilities and China regulations.
- A global supply chain exemplifies a strong "on-the-ground" team in the Ukraine. Recent political issues have caused significant problems in the region. The safety of employees, suppliers, operations, and transportation has been a major issue. Additionally, the government has "tightened" procedures to manage the crisis. But the Ukraine business has managed to meet critical goals because it has a strong "on-the-ground" team. The team has found creative ways to help with safety, security, and governmental regulations. These Ukrainian resources had local

supply chain contacts and a clear understanding of local capability to solve issues 10 times faster than nonlocal corporate resources. Some competitors without these local teams were forced to temporarily stop business during the crisis.

5. Rapid Response (Supply Chain Visibility)

Our fifth best practice in "managing complex, global supply chains" is rapid response systems. Managing supply chain risk is an expensive and critical process in global supply chain management. For this best practice, we have included the following supply chain visibility case study.

Rapid response is a popular and important concept in the supply chain community today. Rapid response is possible only when there is clear visibility across the global supply chain. Global natural disasters in Chile, Southeast Asia, and Japan, combined with political unrest in the Ukraine and the Middle East, have motivated supply chain professionals to review their risk mitigation plans. When managing a more complex global supply chain, this becomes a bigger challenge. A key issue is the time it takes to get back in business or working an alternate option to stay in business after a significant event. Therefore, we have included below a more detailed discussion of rapid response and supply chain visibility.

Everyone who plays the famous Beer Game supply chain simulation discovers one immutable truth: *Supply chain performance breakthroughs result from creating visibility across the supply chain from suppliers to customers.* Without this end-to-end visibility, a firm flies blind, trusting intuition and luck to save the day.

Visibility is extremely important in any domestic supply chain, but it reaches another level in the global environment. A domestic supply chain with lead times of days or weeks can easily stretch to multiple months when placed in a global environment. Countless events routinely disrupt a supply chain over such an extensive time span, particularly in emerging third world economies. A global supply chain faces disruptions resulting from demand spikes, natural disasters, political unrest, strikes, unexpected regulatory issues, port problems, and terrorism.

Tools are available to enhance supply chain visibility. *BT Trace* from BT Global is one example of a leading-edge solution. The critical aspects of global supply chain visibility are discussed below.

- **It All Starts With the Customer**

  Supply chain visibility is a critical need for most companies. It requires *not* a quick fix but instead a major strategy and an implementation plan that could span years. The supply chain visibility strategy should start with a survey of customer needs. Most likely, customers will indicate that they need to know up-to-date and accurate information on the *status of their orders*, and especially information on the *physical location in a global network*.

  That is not all. They also need to know how to better *collaborate with their suppliers*, which implies open sharing of data—lots of data. Many firms also need *traceability* on the outbound flow of goods, and this is especially true for companies who might have to contend with a recall. Examples include food/beverage companies due to contamination, pharmaceutical companies due to drug safety issues, and appliance companies due to product safety hazards. A typical customer needs information such as an accurate order status, the data to effectively collaborate with their suppliers, and traceability, all of which should be incorporated in a supply chain visibility strategy.

- **Visibility Means Managing Big Data**

  Firms must manage a massive amount of data to achieve global supply chain visibility. Data from many disparate sources must be captured and translated into a common format, using complex connectors that route various sources of data through a complex network. Harnessing "Big Data" can create significant value by driving profitability, enhancing productivity, and increasing competitiveness.

  The challenge is daunting. Firms must efficiently collect, access, share, store, and interpret these data, then turn them into actionable intelligence. Equally or more important is the need to manage the data securely to avoid outside breaches by hackers. One trend that arguably helps with this challenge is the increasing access to data and software via a cloud-based system, avoiding the "up front" capital investment, which can be critical for start-ups in emerging economies.

- **Visibility of What?**

  In a global supply chain visibility strategy, firms need to be capable of tracking and tracing more than orders. They also need to monitor inventory by location, transportation assets, and even goods within warehouse operations. This visibility needs to extend upstream to the first and second tier suppliers as well as downstream to customers. As noted, this requires managing large amounts of data from different sources in a secure environment.

Visibility extends to both structured data sources (e.g., ERP systems) and to unstructured data sources such as news feeds, weather information, social media, and more. One company tracks weather data, especially extreme weather events, to identify potential disruptions to its supply chain. The predicted path of a hurricane can give advance warning on potential shipping delays to customers and predict supplier disruptions.

Visibility of Visibility

It is not enough to have accurate, timely data. The data need to be displayed in a way that is easy to access and easy to understand. For example, a manufacturer could construct a "control tower" with multiple large monitors that would not only display data, but also have the capability to be queried. Users could easily interrogate the data with specific questions. Pictured in Figure 2.6 is another example, a control tower from a BT visibility solution, BT Global Trace.

Firms need to see and interrogate the data, and they also need to recognize quickly any out-of-standard situations (exceptions) to enable timely corrective action via alerts sent to PCs or mobile devices. Corporate resources have limited capacity, and can only deal with the exceptions embedded in the maelstrom of data flooding the system. For example, a specific container may be stalled in the port at Hong Kong for longer than normal. This situation calls for an alert to be triggered and quick action to be taken. Firms involved in global commerce need to develop a supply chain event management (SCEM) capability.

Figure 2.6 Benchmark SC Risk systems must be highly visual

With SCEM, a company can monitor, track, and correlate all supply chain events against relevant expected milestones within a business process. The goal of SCEM is to keep all users in the supply chain (from materials suppliers and buyers to warehouse managers and product carriers) informed about activity across the supply chain. As SCEM monitors activity in real time, it can sense problems and respond with alerts, as well as send notifications to appropriate users and partners. Companies such as Oracle, SAP, Manhattan, and BT have SCEM solutions.

Control Towers

"Control Tower" is a term increasingly used to describe an information system that allows a company to monitor and manage its orders and supply chain assets. The challenge is to create an information hub that can capture data from widely varying sources. Generally, firms use a third party to set up a control tower. Control towers are far from ubiquitous. In a *DC Velocity* survey, only 12% of respondents said they were operating a control tower. One of the most important reasons for a control tower is to better manage the supply base during a disruption such as a weather or terrorist event.

Nearly every firm needs a strategy to create world-class supply chain visibility capability. The payback is simply too great to ignore or leave to ad hoc, uncoordinated actions. The global environment is changing rapidly. As ante to stay in the game, firms need an information system that can handle big data in a secure, efficient manner, generate real-time alerts, and display query-able data in a clear, user-friendly format.

Summary

Global supply chains have undergone a series of phase transitions over the ages, from trading supply chains to manufacturing supply chains to the current era of the global, IT-enabled supply chains. These transitions have been fueled by technological innovations.

These innovations have resulted in global economic power moving from Asia to Europe in the middle of the 18th century and then to North America in the latter half of the 20th century. Recent trends suggest that the balance of economic power is either moving back to Asia or at least being leveled across the Americas, Europe, and Asia. Supply chain professionals must have robust best practices to stay ahead of the changes in the dynamic global environment as barriers to free-market capitalism are removed.

In this decade, businesses compete on a global platform. There are no geographic boundaries in the search for revenue and profit. As supply chain leaders, we realize that the best global network decisions are based on business and strategic needs. In essence, *there is no one best answer for our work*. The EPIC framework, along with SCND and the best practices for managing complex supply chains discussed in this chapter, provides valuable tools for supply chain leaders to design and manage a winning global supply chain.

We have included a simple chart to provide macro-direction for global SCND. Based on our research and interviews with our benchmark supply chain companies, Table 2.1 (also referenced at the beginning of this chapter) is a good place to start as you enter the complex world of global supply chain design.

Our research shows that there is no one best answer to the complex analysis and decision-making processes required to develop winning supply chains. Figure 2.7 reminds us of the recent shift from Few/Global to Multiple/Local. But your supply chain must decide which design is right for you.

Are your supply chains designed to win? Take the SCND test in Table 2.5 to find out. Let this tool provide insight into your current supply chain. Your supply chain may have initially provided your business a

Figure 2.7 Sourcing priorities have changed since the 1980s. How will your supply chain make future sourcing decisions?

competitive advantage, but your business and the world are changing at a rapid pace. Has your SCND kept up with the pace of change?

Send copies of this simple self-test to your leadership team members, supply chain analysis leaders, and multifunctional business partners. This can be a great supply chain design team-building exercise.

Table 2.5 How Well is Your SC Designed?—Quick and Easy Assessment Tool

Few Global or Multiple Local

Answer the questions on a 0, 1, 2, 3, 4, 5 scale. Reference descriptions of 0 and 5 scores in the chart.
Apply the questions based on your strategic/business needs and supply chain design/capability.

What is a "0"? "Few Global" Attribute	Questions	Score (0–5)	What is a "5"? "Multiple Local" Attribute
Revenue and volume primarily comes from one region/country (>60%)	1. Is your revenue/ volume distributed evenly across the world?		Revenue and volume is distributed across multiple regions (i.e., Asia, NA, SA, Europe, Africa) and no more than 35% from one region.
Low level of responsiveness required—"drumbeat" processes for new product initiatives, relative flat or predictable demand patterns	2. Does your business require a high level of responsiveness?		High level of responsiveness required. Demand is unpredictable (i.e., urgent customer requests for special shipments or large promotion quantities, rapid response to competitive new product initiatives).
Customer service defects hurt our company, but typically most consumers hold their purchase until our product is available or he/she selects another one of our products. There is limited brand switching.	3. Does your business have a high correlation of customer service defects to lost revenue?		If our product is not on the shelf, the consumer will (>80% of time) purchase a competitive product and if delighted with it, he/she may not return to our brands. Our products have a relatively high profit margin so a lost sale is a significant hurt to our profits.

(Continued)

Table 2.5 (*Continued*)

What is a "0"? "Few Global" Attribute	Questions	Score (0–5)	What is a "5"? "Multiple Local" Attribute
New product initiatives are >80% global in design. Products are designed for global consumers. Limited regional adjustments are made to the product design (typically secondary packaging/customer requirements or local label requirements).	4. Are your products regionally specific? Do you typically launch regionally specific new product initiatives?		Current products and new products (formulas, primary packaging) are designed specifically for each region. There is a low level of global SC scale created from standard products, processes, and/or materials.
Technology (equipment, process, product) is highly complex or proprietary. The technology is protected legally by you and/or your suppliers.	5. Do you have simple technology?		Technology can be readily purchased in market. No legal technology protection is available or required.
High level of SC complexity due to unique equipment/processes, complex materials, extreme number of skus (<$1,000,000 revenue/sku), and/or high sku churn (>30% sku turnover per year). >90% of employees need greater high school level skills.	6. Do you have a low level of supply chain complexity (logistics, technical)? Can you produce with available, local talent?		Low level of complexity (industry norm) and a high level of local talent is available.
SC cost is less than 15% of revenue.	7. Do you have a high total SC cost as a percentage of total business revenue?		SC cost is greater than 60% of revenue.

Table 2.5 (*Continued*)

What is a "0"? "Few Global" Attribute	Questions	Score (0–5)	What is a "5"? "Multiple Local" Attribute
Logistics cost is less than 15% of total SC cost.	8. Is your Logistics cost (transportation, customs, duties, warehousing) a high percentage of your total SC (value chain) cost?		Logistics cost is greater than 65% of total SC cost.
Business prefers a low level of inventory, but will readily increase inventory to lower SC cost, mitigate revenue risk, or solve short term SC issues. Inventory/cash is not a key driver of business shareholder return.	9. Is Inventory Management (managing cash to the lowest levels) of high importance to your business?		Our business requires our best effort to manage inventory to the lowest possible levels. Cost and cash trade offs are analyzed thoroughly to determine the optimal decisions. Business shareholder return depends on strong cash results.
	Total Score		

How did we do?

30 to 45: Supply Chain Network should be designed around local/regional manufacturing, key suppliers, warehousing, and SC technical centers.

15 to 30: Supply Chain network requires a more detail TCO analysis to determine optimal network. The SC network design may be a hybrid (i.e., Mega-Regional supply).

0 to 15: Supply Chain Network should be designed around a few, global manufacturers, key suppliers, warehousing, and SC technical centers.

Transition: Global Supply Chains to Managing Risk in the Global Supply Chain

As mentioned in the previous section, taking supply chains global increases cost, complexity, and risk. Although the first two factors are immediately apparent in the firm's bottom line, it is risk that can be the most difficult aspect to manage.

Risk is, by its very nature, volatile. We have limited knowledge of when port laborers in a foreign nation may go on strike, or if terrorists will attack an important manufacturing district in a key supplier's location. These types of disasters are unwanted and unlikely. But even if managers cannot *predict* risk in their global supply chains, they can still *prepare* for it.

The EPIC framework helps managers assess supply chain viability in international regions. Our own epic assessment suggests no country has perfect scores across the board (in economy, politics, infrastructure, and competence grades). This means that there is a certain amount of risk associated with *any* place you choose to locate your supply chain operations.

The omnipresence of risk cannot preclude managers from implementing global supply chains. Risk is ever-changing and evolving, especially at an international level. Yet there is no harm in preparing for uncertainty. In fact, as our research suggests, there is very much to gain by simply understanding what risks your supply chain *could* be subjected to and putting contingency plans in place should a particular risk (or set of risks) be realized.

Chapter 3 is dedicated to measuring risk in the global supply chain. It takes a holistic view of risk and, in particular, how it can affect supply chain operations. There are survey results and interview quips collected from our extensive risk management research. We offer two case studies for risk management best practices. Then, as a practical application, we provide a specific framework for how risk should be managed. We call it: *identify*, *prioritize*, and *mitigate*.

Due to its global nature and systemic impact on the firm's financial performance, the supply chain arguably faces more risk than other areas of the company. Risk is a fact of life for any supply chain. In spite of that, the vast majority of companies give this topic much less attention than it deserves. This chapter aims to make risk slightly less daunting—and markedly more manageable.

3

Managing Risk in the Global Supply Chain

A Report by the Supply Chain Management Faculty at the University of Tennessee

J. Paul Dittmann, Ph.D., Executive Director,
The Global Supply Chain Institute at the University
of Tennessee Haslam College of Business

Executive Summary

Over the last decade, many companies faced extreme supply chain challenges that stretched their capabilities to the breaking point. The preponderance of natural disasters as well as huge economic swings caused extreme challenges across the supply chain. These challenges have not diminished. Supply chains, which once functioned almost on autopilot, face many dangers today in both the global and the domestic markets.

This paper covers a wide range of risks in the global supply chain and offers strategies and tactics to mitigate those risks. This advice is grounded in research that examined how leading supply chain executives identify, prioritize, and mitigate risk in the supply chain.

The research team distributed a questionnaire across a wide range of companies, including retailers, manufacturers, and service providers. The researchers tabulated data from the responses of over 150 different supply chain executives. In addition, the team completed in-depth, face-to-face interviews with senior executives from six prominent companies. Some findings were surprising. For example, it appears that many supply chain executives have done very little to formally manage supply chain risk—despite recent, unprecedented challenges that necessitate doing so. In particular,

- None of those surveyed used outside expertise in assessing risk for their supply chains
- 66% of firms had risk managers in the legal or compliance functions that virtually ignored supply chain risk
- 90% of firms do not quantify risk when outsourcing production
- 100% of supply chain executives acknowledged insurance as a highly effective risk mitigation tool, but it was neither on their radar screen nor in their purview

Given the magnitude of supply chain risk exposure, this last point is especially perplexing—particularly since insurance providers offer solutions to circumvent, protect against, or ultimately help companies financially recover from many of these risks. Insurance companies possess a multitude of readily available data on supply chain risk. Such data can be invaluable in assessing and managing supply chain risk.

When asked about risk, one interviewee had a telling response: "Frankly, my boss is not asking me to look at it [risk]. It [risk management] is the right thing to do, but we are not rewarded for doing it." Maybe that is at the heart of the problem: few executives are compensated or incentivized in their day-to-day job to rigorously manage risks.

This chapter examines these findings and more. But more importantly, it proposes a supply chain risk management process for companies of all sizes. From this analysis we developed Figure 3.1, which illustrates the three-step process to protect your business from supply chain risk.

Figure 3.1 GSCI model for SC risk management

Risk in the Global Supply Chain
Introduction

Due to its global nature and systemic impact on the firm's financial performance, the supply chain arguably faces more risk than other areas of the company. Risk is a fact of life for any supply chain, whether it is dealing with quality and safety challenges, supply shortages, legal issues, security problems, regulatory and environmental compliance, weather and natural disasters, or terrorism. There is always some element of risk.

Companies with global supply chains face additional risks, including (but not limited to) longer lead times, supply disruptions caused by global customs, foreign regulations and port congestion, political and/or economic instability in a source country, and changes in economics (such as exchange rates).

The scope and reach of the supply chain cries out for a formal, documented process to manage risk. But without a crisis to motivate action, risk planning often falls to the bottom of the priority list. The low priority for managing risk in companies is puzzling. After all, supply chain risk management is a very popular topic at conferences and is written about extensively in books and articles. However, in spite of all of the discussion, we still see the vast majority of companies giving this topic much less attention than it deserves.

Supply chain executives drive this risk apathy, though not by design. These executives often find themselves at the center of the daily storm, striving to balance very demanding operational objectives with satisfying customers, cutting costs, and growing revenue. They must deliver results today while working on capabilities that will make their companies competitive in the future. They operate in the same maelstrom of competing priorities and limited time as their executive peers—but their scope of activities is broader, and they have less direct control over all the moving parts. In this environment, risk management receives a much lower priority than it should.

But the repercussions of supply chain disruptions to the financial health of a company can be far-reaching and devastating. A study by Hendricks and Singhal[1] emphasizes the negative consequences

---

[1] K. Hendricks and V. Singhal, "An Empirical Analysis of the Effect of Supply Chain Disruptions on Long-Run Stock Price Performance and Equity Risk of the Firm," *Production and Operations Management* 14, no. 1 (2005): 35–52.

of supply chain disruptions. The study analyzed over 800 supply chain disruptions that took place between 1989 and 2000. Firms that experienced major supply chain disruptions saw the following consequences:

- Sales were down 93% (over a 3-year period)
- Shareholder returns were 33% to 40% lower (over a 3-year period)
- Share price volatility was 13.5% higher
- Operating income declined by 107%
- Return on Assets (ROA) declined by 114%

But there is a silver lining. While risk cannot be eradicated, it can be identified, assessed, quantified, and mitigated. Once a risk management plan is developed, it can become a competitive advantage because so few firms have one.

Risk: A Daily Fact of Life

Supply chain professionals from very large to very small companies face a wide range of risks that never make the headlines. Indeed, the Japanese tsunami and earthquake riveted the world a few years ago, but in the meantime, supply chain professionals have to deal with the unexpected day-to-day challenges that are less public, but just as disruptive.

Supply chain experts at UPS Capital, who specialize in risk mitigation, divide supply chain vulnerabilities into two categories of risk. The first are day-to-day risks, or risks provoked by the normal challenges of doing business. These risks include:

- Customer demand changes
- Unexpected transit delays
- Supplier problems that delay critically needed components
- Theft, a much larger problem than most realize
- Production problems
- Warehouse shortages that cause serious delays in customer shipments
- Cyber security

The second type of disruption is the one where "all hell breaks loose." Risks like these—epidemics, tsunamis, and terrorism—usually cannot be predicted, but companies should have a *risk management process* to mitigate and minimize the impact of such events.

Although this paper more easily references the higher profile problems, the day-to-day problems cannot be ignored. Later in this paper we discuss methods to assess, quantify, and mitigate supply chain risks, whether large or small, routine or extraordinary, forecasted or unexpected.

Insurance: A Surprising Finding

Although we discuss risk mitigation strategies later in the paper, we thought it important to highlight one of the most telling findings here. We were surprised to learn that insurance is simply not on the radar screen of supply chain professionals as a risk mitigation approach. As it happens, when we discuss the usefulness of insurance with these professionals in interviews, they quickly realize they have missed a highly effective tool.

Insurance companies and brokers are eager to share best practices because they, too, have a vested interest in avoiding losses. They can be key partners in working with firms to minimize the financial effects of daily supply chain risks and catastrophic disruptions once the loss occurs. But, more importantly, they can help companies find solutions to prevent the day-to-day problems that result in losses, thus avoiding the disruption and the subsequent claim settlement. No one wins in a loss. Insurance companies regularly see the best and worst of supply chain practices and need to be on the winning side of mitigating risk for their clients—and their own bottom lines.

That is why specialized providers, including logistics companies, have entered the market with products specifically designed to mitigate supply chain risk. With volumes of logistics data, years of industry experience, and proprietary visibility tools, these companies offer new risk mitigation solutions that traditional business owners' policies do not provide.

For example, one recently introduced service was designed for the health care industry. Proprietary technology proactively monitors expensive and highly sensitive shipments for time and temperature requirements. If the shipment is in jeopardy, proactive measures (such as re-icing or expediting to same day delivery) are taken. If the shipment is lost, damaged, or delayed beyond the point of recovery, the insurance company reimburses the customer for the full sales value rather than just the cost of the goods. Insurance can be about much more than receiving payment. The right insurance experts can help businesses avoid risk.

Specialized insurance services that come from diverse insurance industry providers can be an integral component of a company's risk mitigation approach. Before considering insurance or other risk mitigation solutions, most companies should consider and attempt to quantify the risks they face. Yet, as described later in this report, few companies formally undertake this critical first step.

The Alarming State of Supply Chain Risk Management

Despite how important it is to identify and mitigate risk as part of supply chain strategy, our research rarely found robust risk practices among firms that pursue a global outsourcing strategy. We have found that when companies analyze highly risky global outsourcing decisions, they fall into three categories based on the assessments they conduct:

- Category 1 (36%): consider only unit cost and transportation
- Category 2 (54%): consider unit cost, transportation, and inventory
- Category 3 (10%): consider unit cost, transportation, inventory, and a risk quantification and assessment

In other words, 90% of firms *do not formally quantify* risk when sourcing production. As one Senior Vice President of supply chain told us in an interview, "On paper, and without the 'risk' thing, this global sourcing deal looks like a great return on investment. With risk, who knows?"

In a study done by *Risk and Insurance Magazine*, none of the 110 respondents rated their company as "highly effective" at supply chain risk management.[2] Two-thirds described their effectiveness as low or "do not know." The typical supply chain manager estimates that just 25% of his company's end-to-end supply chain is being assessed in any way for risk.

Few supply chain professionals would dispute that the supply chain strategy in their firms should identify possible global supply chain risks, develop probability and impact assessments, and then

---

[2]"Supply Chain Survey Report," *Marsh, Inc., and Risk and Insurance Magazine*, April 15, 2008.

create risk mitigation plans. Executing this process can help avoid much pain later. Practical examples to address these opportunities are included later in the chapter.

An Up-to-Date Twist on Risk: The Survey Says . . .

Given the popularity of supply chain risk at conferences and for articles, one could reasonably ask, "Is there anything new to say on the topic?" We definitely think so. The intelligence on this subject gets increasingly sophisticated, driven by the continuing complexities of the global environment. As a foundation for this chapter, we conducted a major survey of the state of supply chain risk management today. We obtained input from over 150 supply chain executives across multiple industries, and we conducted in-depth face-to-face meetings with six companies.

The survey results provide the basis for this chapter and allow us to put an up-to-date "twist" on risk. Our intent is to provide practical guidelines that companies can use today to mitigate and manage supply chain risk.

The survey data allow us to make modern observations on the following five topics:

1. Documented risk management processes
2. Facility loss and backup plans
3. Supplier loss and backup plans
4. Supply chain risks
5. Risk mitigation strategies

These five topics are discussed in detail in the following pages.

1. Documented Risk Management Processes

None of those surveyed uses outside expertise in assessing risk for their supply chain. Instead, virtually all (93%) soldier on, doing the best they can within their own departments. The rest admit they do not consider risk at all. Figure 3.2 illustrates the disparity between those who assess risk internally, and those who do not assess risk at all.

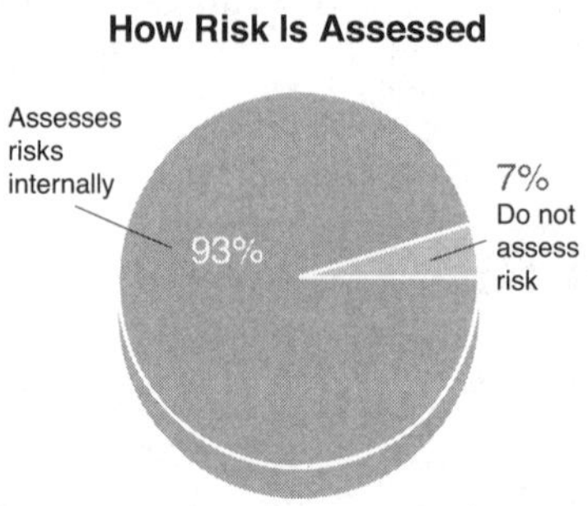

How Risk Is Assessed

Assesses
risks
internally

93%

7%
Do not
assess
risk

Figure 3.2 SC Risk is an external process
and is typically assessed internally

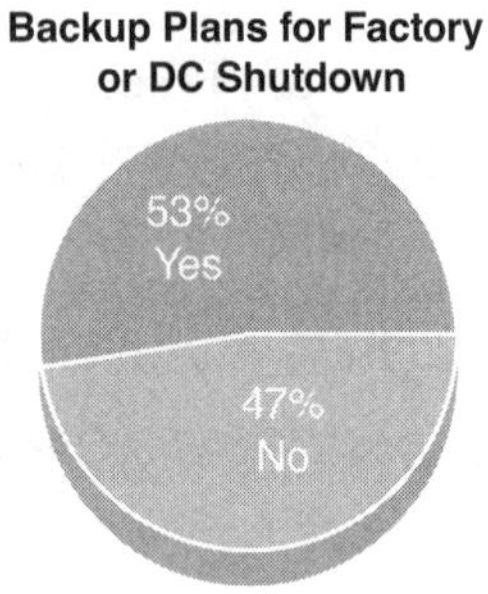

**Backup Plans for Factory
or DC Shutdown**

53%
Yes

47%
No

Figure 3.3 Most supply chains do
not have a "back up" plan

The majority (66%) of companies has a risk manager somewhere in the firm, often in the legal or finance areas. But almost all of these internal company risk assessments ignore supply chain risk. Instead, they focus on product liability or overall financial issues that could impact shareholder value in a material and very public manner.

2. Facility Loss

If a natural disaster or major equipment failure shuts down a company facility (such as a factory or a distribution center), about half of the firms surveyed (53%) have a backup plan that can be implemented fairly quickly. The bad news is that the other half (47%) *does not* have a backup plan. Figure 3.3 shows this split in a graphical format.

If disaster strikes, about 7 in 10 companies (69%) have a documented response plan in place to salvage business with their customers through product substitution, proactive communications, or inventory. This means that almost a third of companies do not have any disaster response plan in place for supply chain risk. You will see in Figure 3.4 that 31% of companies fall in this category.

3. Supplier Loss

The survey, visualized in Figure 3.5, found that nearly half (45%) of supplier spending for United States–based companies is outside the United States, with 20% in Asia. Of course, longer supply lines increase supply chain risk.

Documented Response
Plan to Serve Customers

Figure 3.4 Most supply chains do not have a plan to respond to a crisis

On average, about 49% of the firms surveyed had suppliers who could continue to supply if they suffered a disaster in one location, meaning that over half (51%) *could not* continue supplying within a reasonable time frame. The bar chart in Figure 3.6 shows these percentages in graphical format.

As you can see in Figure 3.7, firms vary widely in terms of how many of their suppliers are sole sourced. In this survey, 38% of suppliers are sole sourced. But the spread is very broad. At just one standard deviation, the range for sole sourcing among the firms surveyed was 13% to 63%. It can safely be said that many firms take on the risk of sole sourcing with a relatively large number of their suppliers. Some do this for economic reasons, such as when one supplier has a significantly lower cost and/or higher quality, while others have practical reasons, such as when no other supplier can adequately satisfy the

Global Spending by Region

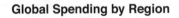
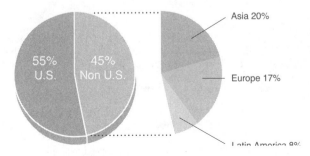

Figure 3.5 Asia is a key priority for SC risk management

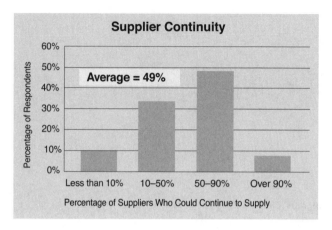

Figure 3.6 Only half of our suppliers have SC risk plans

company's needs. Still others, unfortunately, may do this for relationship reasons, citing "we have always done business this way."

The pie chart in Figure 3.8 shows that the vast majority (86%) of companies are multiple sourced with their domestic and global transportation carriers. Very few companies (14%) single source with transportation carriers, and less than a third of those have any concern about the sole sourcing arrangement.

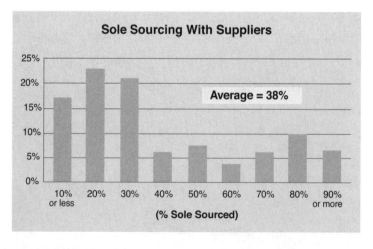

Figure 3.7 Sole sourcing can be the best strategy but requires robust risk management

**Transportation
Carrier Sourcing**

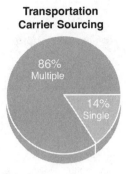

86%
Multiple

14%
Single

Figure 3.8 Transportation is
one SC discipline with robust risk
management

**Global Shipment
Concentrations**

75%
Known &
Comfortable

18%
Unknown

7%
Known &
Uncomfortable

Figure 3.9 Most SC leaders are concerned
with concentration of global shipments

Eighteen percent of those surveyed do not know the degree of concentration of their global shipments, and 7% say they know the degree of concentration and are uncomfortable with it. On the other hand, a large majority (75%) tracks this information in order to comfortably manage risk in global shipping (see Figure 3.9).

4. Supply Chain Risk Ratings

From our extensive research and analysis, we were able to chart in Figure 3.10 just how concerned supply chain professionals were with specific risks. The *number one risk* on the minds of those surveyed was potential quality problems. Long global supply lines make it very difficult to recover from quality issues. For example, Whirlpool decided years ago to outsource the production of dishwasher water seals to a Chinese supplier for a net savings of $0.75 per unit. This totaled over $2 million in annual savings. But soon after the arrangement was made, the Chinese supplier changed to a different rubber supplier. The seals made from this new rubber leaked in dry climates, causing a failure rate of nearly 10%.

By the time Whirlpool discovered the problem, over two million dishwashers had been produced with the defective seal, and two months' worth of supply was in transit on the ocean. This cost the company millions of dollars and destroyed all savings from the project for over three years. Whirlpool could have avoided this problem by doing more planning and putting robust quality controls in place.

Figure 3.10 Interviews show SC leaders are concerned about a broad range of SC risks

Those controls now exist at Whirlpool and are excellent for the company as a whole, but it took a crisis to fully motivate action.[3]

The *number two risk* concerns the requirement for increased inventory due to a longer global supply chain. Twenty years ago, companies were so caught up in the allure of low-cost labor that they rushed their operations to Asia. This extended their supply chains, eventually requiring them to carry at least 60 to 75 days of supply in additional inventory. Most firms now understand the stress this additional inventory places on their company's working capital and cash flow. When contending with extremely long global supply lines, supply chain professionals find it extremely difficult to achieve the aggressive inventory turnover goals given to them by the CEO/CFO.

Both supply chain professionals and CFOs agree that burdening a company's working capital with the cost of added inventory requires a focused management effort to minimize the impact. Most supply chain professionals are not familiar with the available financial

---

[3]R. Slone, J. Mentzer, and J. Dittmann, *The New Supply Chain Agenda: The Five Steps That Drive Real Value* (Boston, MA: Harvard Business School Publishing Corporation, 2010).

products that allow companies to maximize the amount of cash that can be borrowed against trading assets, such as inventory that is warehoused internationally and inventory that is in transit.

Financing products can be especially helpful to growing companies that need to maximize cash flow. A properly structured working capital loan or other financial bridge can help a company minimize financial strain and work its way through identified risks. For example, one company had a 270-day cycle time from its Chinese supplier, tying up cash in inventory for nearly nine months. By assessing the situation and quantifying the need for additional cash flow, the company used its in-transit inventory as collateral for a loan, thus preserving cash flow.

Natural disasters stand as the *number three–ranked risk*, no doubt brought to top of mind with recent high-profile natural disasters such as the Japanese tsunami and Thailand flooding.

Of least concern to supply chain professionals is terrorism and piracy, followed by customs delays. Firms have gotten much savvier about dealing with customs issues. They are taking full advantage of programs that speed customs processing, such as C-TPAT.

5. Risk Mitigation Strategies

The number one strategy used to mitigate supply chain risk is to choose financially strong, competent, world-class suppliers. That is easier said than done. Firms tell us that it takes approximately two years to develop and fully certify a global supplier.

The second strategy used to mitigate supply chain risk focuses on compressing global shipping and cycle time variations. Leading firms apply Lean principles and Six Sigma techniques to this effort. They map the value stream of the end-to-end global shipping process and look for ways to reduce or eliminate waste and delays at every step.

The third-ranked strategy used to mitigate supply chain risk involves the use of visibility tools to closely track global shipments and take action when necessary. Leading firms use supply chain event management technology to send alerts to key personnel when action needs to be taken by someone, somewhere in the global supply chain to address potential delays.

Other observations from the survey data include:

- Predictive modeling, arguably the most sophisticated risk mitigation technique, seems to be as popular as merely reacting to a crisis with air freight or expedited shipping, which is the least sophisticated risk mitigation technique.

- At number eight, near-shoring is well down the list as a risk mitigation approach. Even though the trend to outsource globally is slowing, it does not mean there is a rush back to the United States.

Failure Mode and Effect Analysis

A company cannot devote enough resources to mitigate all risks. It must have an approach in place to identify the most important ones first. A great method for doing that is the failure mode and effect analysis (FMEA) approach. The military first used the FMEA approach as far back as the 1940s. It prioritizes risks based on three factors:

- Seriousness of consequences
- Likelihood of the problem ever occurring or frequency of occurrence
- Likelihood of early detection of the problem

Several firms have successfully applied this approach as a way of identifying high-priority risks to the supply chain. This allows them to determine which risks require a mitigation plan and which are too low impact or unlikely to warrant the effort. The real power of this approach lies in its use as a framework to discuss and debate risks with the supply chain strategy team. Given that risk analysis has a large subjective component, reaching consensus is critical. Two examples of this approach are described later in the chapter. A company cannot devote enough resources to mitigate all risks, but Figure 3.11 shows which risk mitigation efforts are the most preferred.

Using insurance as a risk mitigation tool is ranked last. We believe this is a lost opportunity for supply chain professionals, as we discussed at the beginning of this section and as we will discuss later in this paper.

Best Practice Case Studies in Supply Chain Risk Management

While doing the research for this chapter, we encountered two companies that are leaders in supply chain risk management: IBM and Lockheed Martin. Although these are very large companies,

Figure 3.11 Priority setting is key for a robust risk management plan

we believe their stories hold valuable lessons for firms of any size. In addition, we found an outstanding supply chain risk management tool called Supply Chain Risk Identification Structure (SCRIS) and another called the Risk Exposure Index. The IBM and Lockheed Martin case studies, as well as the SCRIS and Risk Exposure Index tools, are described below.

IBM

IBM's chief risk management officer's assignment is to implement and sustain an enterprise risk management process. The goal is to ensure a world-class risk management process for each business unit. With IBM's huge global outsourcing budget totaling tens of billions of dollars, and a network of over 20,000 suppliers, its supply chain is complex—especially since some suppliers are, by necessity, sole-source suppliers. Executing a global sourcing strategy where sourcing is conducted across developing countries (and with more than just first-tier suppliers) has a cumulative effect on the amount of risk introduced into the supply chain.

To manage risk, IBM's global sourcing process looks far beyond unit cost to the total cost picture. All of the dependencies are fully mapped, and whenever possible, backup sources are specified. The top risks are identified, along with their impact on the company's supply chain. IBM also develops contingency plans for events that will inevitably happen at some point in the future (e.g., a pandemic).

Several years ago, IBM developed and patented its Total Risk Analysis (TRA) tool. The need for this tool arose from the tremendous complexity of its supply chain, whose interactions went far beyond the ability of spreadsheets to comprehend. Initially, IBM assumed it could purchase a tool to do this, but the company quickly found that an acceptable option simply did not exist. The existing tools focused heavily on financial data modeling and fell far short of a comprehensive supply chain risk analysis.

IBM developed its TRA tool to collect a multitude of data on many dimensions from 53 countries. Countries are further divided into logical economic entities. The tool filters risks into a critical few, showing high-risk country/product/component combinations. Since it is important to avoid overwhelming the line organization, only the most important risks are surfaced for a required mitigation plan.

In March 2011, the Japanese earthquake and tsunami tested the TRA tool, and it responded flawlessly. Within a few hours of the disaster, IBM used the tool to determine all of its potential supplier problems. It was then able to immediately assemble details and develop backup plans.

The IBM risk management system goes far beyond the TRA tool. It encompasses an entire management system with business process disciplines. Although TRA is at the core, it is the management processes in place that allow for effective use the tool and the ability to react appropriately.

Mike Ray, IBM's Vice President of Business Integration and Transformation, spoke at the University of Tennessee's Supply Chain Forum in November 2013, where he told a story that demonstrates IBM's ability to react to a supply chain crisis. On July 25, 2008, just after an IBM executive embarked on a train in New York City, he received an instant message. The alarming message read, "There have been a series of bombings in Bangalore, India, earlier today!" The IBM sourcing center in Bangalore issues a huge volume of purchase orders for thousands of suppliers around the world, and IBM (as well as those suppliers) absolutely depends on it. Its shutdown could mean a major disruption in IBM's supply chain.

As the train sped into the city, Mike messaged back that he hoped all the people in Bangalore were safe, and that anyone not working should stay home. He then quickly determined that only 25% of the center's operations were feasible without a certain percentage of available employees. By 7 a.m. EST, while still on the train, he transitioned the remaining operations to a sourcing center in Budapest. By 7:30 a.m., arrangements had been made for the rest of the operations to be picked up by a center in Endicott, New York. In just one hour, IBM created duplicate support coverage, resulting in no supply chain disruptions anywhere in the world. When the IBM executive stepped off the train at 8 a.m., everything had been resolved.

Lockheed Martin

As with IBM, Lockheed Martin's Aeronautics Company (LM Aeronautics) supply chain risk management tool instantly indicates all of its supplier dependencies. If a disaster hits, the company immediately knows where to focus efforts to assess the impact. A company like Lockheed Martin has more to worry about than most because of its wide range of supply chain risks, especially its intellectual property risks.

The LM Aeronautics process focuses on risk mitigation with plans required for high-priority risks. The company has developed a set of risk mitigation resources, which specify a list of questions to address common risk events. They also prescribe established lessons learned and best practices. The process directs "boots on the ground" where appropriate, and engineering design standards consider complexity as a means of reducing risk.

As part of the disciplined risk management process, the company periodically conducts risk surveys at each major supplier site. Risks are identified, and detailed mitigation plans are developed. Finally, LM Aeronautics tracks and scorecards the completion of risk mitigation plans.

The risk management process also includes an analysis of suppliers' financial health, dashboards for suppliers, current and future demand on the suppliers, and capacity modeling, especially in light of the demands of other aerospace manufacturers (e.g., the demand for commercial aircraft can consume much of the capacity for certain commodities in the supply chain).

The Supply Chain Risk identification Structure (SCRIS)

This tool was developed by the Council of Supply Chain Management Professionals (CSCMP) and Competitive Insights LLC to provide a reference model for identifying, mitigating, and measuring supply chain risk. SCRIS takes overall business continuity risk and breaks it down into multiple categories and multiple tiers. It provides an excellent framework and checklist to manage overall supply chain risk. SCRIS can be used to develop supply chain risk management strategies. Just as important, it facilitates communications across the organization and motivates the appropriate level of focus on supply chain risk management.

Risk Exposure Index

The Risk Exposure Index methodology was created by David Simchi-Levi at the Massachusetts Institute of Technology (MIT). The index looks at a firm's entire supply chain, including first- and second-tier suppliers. It then estimates a "time to recovery," or TTR, to full functionality after a major disruption. Once the TTRs are known for each node in the network, it is possible to compute the financial impact on the firm and prioritize accordingly. This methodology invariably leads to actions to reduce the TTR for critical nodes in the network.

Recommendations to Manage Supply Chain Risk

To manage supply chain risk in your company, you will have to develop processes to identify, prioritize, and mitigate risk.

- **Risk identification**
  o What can go wrong?

- **Risk assessment**
  o What is the likelihood it will go wrong?
  o What is the magnitude of the consequences and overall impact on the firm?
  o How quickly will the problem be discovered?

- **Risk mitigation and management**
  - o What options are available to mitigate the risks?
  - o What are the costs and benefits of each option?

Leading companies such as IBM and Lockheed Martin have a process that executes these steps continually over time. In the dynamic global environment, change is a constant. Risks identified and mitigated today become obsolete tomorrow. Risk management must be a dynamic, ongoing process.

Identify Risks

Your supply chain strategy team should set aside time to evaluate the risks facing your supply chain. As one supply chain professional told us after experiencing his share of costly risk situations, "Intuition and gut feel will not cut it. You must have a disciplined risk management process." Risks come into focus once you have determined your customers' needs, completed an internal best practice evaluation, and assessed competition and technology. You should be in a good position to do that, assuming you have established a foundation to place those risks in perspective.

You will have to gather data on your suppliers and on the countries in which you do business. And, you will have to create a way to organize the data, perhaps leveraging tools like the SCRIS and Risk Exposure Index.

You will then have to schedule time for your team, perhaps assisted by outside experts, to brainstorm all the potential risks faced by your supply chain. This should be a free-flowing meeting, with plenty of time set aside. Members of the group should not let themselves feel intimidated or overwhelmed. The idea is to get everything on the table without any constraints or criticism. The team would ideally get together in a one-day off-site meeting to identify the risks facing the firm's global supply chain. Listed below are some risks your supply chain may face:

- *Routine supply chain risks*: These involve events like unexpected transit delays, changes in customer orders, problems with suppliers, theft of property, and warehouse/production malfunctions, all of which can cause serious delays in customer shipments.
- *Natural disasters*: Although these are unpredictable, a few firms try to anticipate climatic disruptions and develop contingency plans. If a company has a facility in a hurricane-prone area, it

can assume it is only a matter of time before the odds catch up with the location.

- *Quality problems*: A long supply line often exacerbates quality issues. This risk causes companies to carry more inventory. One consumer durable products manufacturer discovered a quality problem and was mortified to find that it had two months' worth of supply in transit on the Pacific Ocean, all with the same defect.

- *Forecast error*: Long-range forecasts required by long global supply lines are notoriously inaccurate. Forecast error over long global lead times often results in major availability issues and excess inventory problems.

- *Damage*: Whether you are importing or exporting, there is significantly more handling in the supply chain that exponentially increases the chance for damage.

- *Political/civil unrest*: While not a major concern, it should be on a company's risk list and examined, depending on the countries of import and export.

- *Culture clash*: In 2010, 18 people at Foxconn (the company best known for manufacturing Apple products) attempted suicide over fears of job cuts.[4]

- *Strikes*: Strikes are a reality and must be anticipated. The 40-day Hong Kong port workers strike in April and May of 2013 caused numerous supply chain disruptions. Strikes could also occur at production plants or facilities that supply critical parts.

- *Laws and regulations*: Unusual or unexpected application of regulations in a particular country must be considered, as must the Foreign Corrupt Practices Act in the United States.

- *Customs or port issues*: Customs regulations are always in flux. Failure to understand the rules and regulations can often cause excessive shipment delays and fines.

- *Terrorism*: Although quite rare, acts of terrorism often result in the addition of permanent costs to the supply chain far beyond the cost of the act itself.

---

[4]M. Heffernan, "What Happened After the Foxconn Suicides," *CBS Moneywatch*, August 7, 2013, http://www.cbsnews.com/news/what-happened-after-the-foxconn-suicides/

- *Safety problems*: How many times are safety recalls issued on top-name brands? There may be opportunities with product liability insurance to mitigate risks, from product design to manufacturing.

- *Changes in economics*: For example, wages in China are escalating for a variety of reasons. Some point to the "one baby" policy as a source of future increasing labor shortages, even though predictions call for that policy to relax. As reported by China's Labor Bulletin, under China's current Five-Year-Plan (2011–2015), the minimum wage is slated to increase at an average rate of 13% a year.[5]

- *Price or currency fluctuations*: Extreme and unexpected changes in the price and availability of critical raw materials wreak havoc on a firm's financial plans, as do swings in currency.

- *Intellectual property loss*: This is a major problem that should not be underestimated. Many firms, to their chagrin, have found they inadvertently created a new global competitor because they were unprepared for the theft of intellectual property.

- *Siloed business processes*: For example, marketing can initiate a major promotion event that drives a spike in demand without allowing the global supply chain to plan ahead.

- *Technology*: Failed implementation of supply chain technology, such as Wal-Mart's RFID saga, can have a huge negative impact on the supply chain. The six-year RFID drama may have moved the technology along a bit faster. But many believe it was extremely premature.

- *Pirate attacks*: Piracy on the world's seas recently reached a five-year low, although it is still a danger, with 297 ships attacked in 2012, compared with 439 in 2011.[6]

- *Third-party risk*: The way your suppliers do business could unexpectedly impact your firm in a devastating way. This requires more in-depth discussion in the following section.

---

[5]"Wages in China," *China Labour Bulletin*, June 10, 2013, http://www.clb.org .hk/en/content/wages-china

[6]"Piracy," *World Shipping Council*, 2015, http://www.worldshipping.org/industry -issues/security/piracy

Third-Party Risk

You should be extremely sensitive to the manner in which your global suppliers do business. For example, on April 24, 2013, an eight-story garment factory in Bangladesh collapsed, killing scores of workers.[7] This building served several prominent retailers.

Shortly after this event, many US retailers announced plans to improve factory safety in developing countries.[8] The bad publicity was exceeded only by the human toll in this tragedy, as a number of retailers faced intense and negative public scrutiny.

More firms realize that selection of third-party suppliers and the way those suppliers do business can have a huge impact on their organization. Therefore, more companies are putting processes in place to ensure their suppliers are:

- Doing business without any hint of corruption (including bribery)
- Handling data in a secure manner
- Managing competently to maintain financial solvency
- Complying with the laws of the countries in which they operate, and
- Enjoying a good reputation where they operate by doing business legally and morally

Supply chain professionals know that they have to be vigilant all over the world: China, India, Africa, Russia, South America, etc. Each of these regions has its own customs, laws, and practices—both visible and hidden.

Companies discover that supplier reputation trumps cost. This brings the vetting of global suppliers front and center. Leading companies rank their suppliers based on the potential risk they represent. Some use outside investigators. They develop checklists that the sourcing department must document and check off as they proceed through the procurement process. They perform annual audits and/or make unannounced visits. Due diligence is never done.

---

[7] J. Manik and J. Yardley, "Building Collapse in Bangladesh Leaves Scores Dead," *The New York Times*, April 24, 2013, http://www.nytimes.com/2013/04/25/world/asia/bangladesh-building-collapse.html?_r=0

[8] M. Moylan, "Retailers, Including Target, Grapple With Unsafe Foreign Factories," *MPRnews*, August 5, 2013, http://www.mprnews.org/story/2013/08/05/business/target-developing-nation-unsafe-factories

Other Approaches to Identifying Supply Chain Risk

There are other activities that your team can use in a one-day off-site meeting to tease out potential risks. Often a valuable exercise is to engage in "war games" simulations. For example, what would happen if a major port were shut down for an extended period of time due to a catastrophic event like a dirty bomb explosion?

As part of a strategic planning process, your firm should invest time to consider disaster scenarios. Although you cannot know when such events are going to occur, you can anticipate possible disruptions based on the location, geography, climate, and political environments of the places goods are sourced. Then, engage in mock exercises to prepare your company to better react to supply chain disasters. As an outcome of these exercises, your team should develop a risk management process that should be followed in the event of a crisis. An important part of this process includes a plan for communicating quickly to all stakeholders.

A few high-profile examples may help your team in its disaster contingency planning. In August 2007, Mattel had to recall 19 million toys manufactured in China because of lead paint issues. Hasbro's stock surged and Mattel's fell, and the gap in shareholder value remained pronounced for over two years.[9]

Other high-profile examples include natural disasters like the 2010 Eyjafjallajökull volcano eruption in Iceland. Or, the twin 2011 global disasters: the 8.9 earthquake and resulting tsunami in Japan and the massive floods in Thailand. These events were perhaps somewhat predictable in the sense that Japan is a volcanic region and Thailand is prone to flooding, but the scale of the 2011 events proved more extreme than even the most aggressive risk managers could have imagined.

Disasters such as these only occur every 100 to 200 years. But because so many events can happen in a globally interconnected supply chain, the probability is high that something major will impact your supply chain.

Sourcing offshore clearly carries a wide range of risks as evident in the examples above. Long supply chains offer more opportunities for

---

[9]D. Barboza and L. Story, "Mattel Recalls 19 Million Toys Sent From China," *The New York Times*, August 15, 2007, http://www.nytimes.com/2007/08/15/business/worldbusiness/15imports.html?pagewanted=all

disruption by unforeseen events. Because of the impact on the corporation, global supply chain strategies must include a thorough risk analysis. But any impression that supply chain risk is an exclusively global phenomenon should be quickly dispelled. In 2013, for example, there were 128 natural disasters (severe weather, floods, earthquakes, and fires) in the United States alone. These events resulted in over $21 billion in business losses, with over 120 people losing their lives.[10] A *Supply Chain Digest* article in 2006 detailed the "eleven greatest supply chain disasters of all time." All of these events took place within the borders of the United States, and most had nothing to do with natural disasters.[11]

Countless other examples exist of local supply chain disasters and their devastating impact on the firm's performance, ranging from natural disasters to software failures. Whether local or global, supply chain risk must be identified, prioritized, and mitigated.

Prioritize Risks

Once your team has identified the risks facing your supply chain, the next task is a daunting one: prioritization. Line organizations have the capacity to deal seriously with only a few high-priority supply chain risks, making prioritization crucial in the risk management process. A couple of examples showing how this has been done in other companies may be helpful. The examples rely on a version of a process engineered long ago to identify and prioritize risks, as well as the FMEA approach mentioned earlier in the paper.

Example One: Addressing Supply Chain Risk at a Food Manufacturer Using FMEA

As part of its supply chain strategy, a food products manufacturer was considering outsourcing warehouse operations to a third party. To assess the risks associated with this move, the manufacturer used a table (much like Table 3.1) to guide the risk discussion.

---

[10]"2013 Natural Catastrophe Year in Review," *Munich RE*, January 7, 2014, https://www.munichreamerica.com/site/mram/get/documents_E1433556406/mram/assetpool.mr_america/PDFs/4_Events/MunichRe_III_NatCatWebinar_012014.pdf

[11]D. Gilmore, "Worst Supply Chain Disasters," *Supply Chain Digest*, January 26, 2006, http://www.scdigest.com/assets/FirstThoughts/06-01-26.cfm (retrieved January 23, 2011)

Table 3.1 Using Probability Analysis to Help Prioritize Risk Plans

Food Manufacturer Risk Analysis°		
	Risk 1: Safety of Food Product	Risk 2: Freshness of Product
Severity°° (1–10)	9	6
Probability of Occurrence°° (1–10) High probability − 10 Low probability − 1	2	4
Probability of Early Detection°° (1–10) High probability − 1 Low probability − 10	6	2
Probability Index (Multiply Three Items Above)	9 × 2 × 6 − 108	6 × 4 × 2 − 48
Recommended Action	Purchase insurance	Audit inventory and ensure stock rotation
Responsibility	Safety engineering	Third party with company oversight

°Figure developed by the University of Tennessee for classroom instruction. Later used in the book *Supply Chain Transformation*, 2012, J. Paul Dittmann, page 134

°°Scores determined by group consensus

The supply chain group identified 13 risks and, using the approach outlined, prioritize them. Eventually, the group decided to launch mitigation projects for the top five prioritized risks. Table 3.1 illustrates using just two of the risks identified.

After using the FMEA process to prioritize the various risks, the company established a risk mitigation plan (called "recommended action" in Table 3.1). Once the supply chain group completed and gained approval for the supply chain strategy, it assigned responsibility for each of the five high-priority risk management projects and made those projects a standard part of the weekly strategy implementation meeting.

Example Two: Addressing Supply Chain Risk at a Durable Goods Manufacturer

In another example, a durable goods manufacturing company felt that its supply chain strategy should include some important outsourcing decisions. It decided to test that assumption by evaluating the risk associated with outsourcing a key manufacturing component to a Vietnamese manufacturer. The manufacturer used a modified version of the previous approach, but this firm focused on two factors. First, the group estimated (through consensus) the *probability* of occurrence for each risk and then multiplied that by the *estimated cost* of the occurrence. The analysis looked very much like the one shown in Table 3.2.

The firm then used this analysis in two ways:

1. The supply chain team made sure the ROI on the outsourcing project included the "cost of risk," which in this case was $22.12 per unit. The outsourcing savings without risk stood at a net $55 per unit, which safely exceeded the risk estimate. The group fully recognized the subjective nature of this analysis. Yet, the act of discussing the potential sources of disruption and estimating their costs gave the team some assurance that the project would still be viable even if it encountered one or more of the potential risks.

Table 3.2 Using Probability Helps to Assign a Cost to Risks

	Outsourcing Risk Analysis*		
Risk	Estimated Potential Loss, Stated as Cost in $ Per Unit	Subjective Probability of Occurrence	Net Loss Per Unit (Multiply the Prior Two Columns Together)
Quality Failure	$25.00	0.10	$2.50
Safety Failure	$100.00	0.01	$1.00
Unexpected Demand Spike	$30.00	0.25	$7.50
Currency Change	$20.00	0.25	$5.00
Intellectual Property Problem	$10.00	0.25	$2.50
Source Disruption Force Majeure	$30.00	0.10	$3.00
Port Problem	$25.00	0.025	$0.63
		Total	$22.13

*Figure developed at the University of Tennessee for classroom instruction. Later used in the book *Supply Chain Transformation*, 2012, J. Paul Dittmann, page 135

2. Once the supply chain team completed the supply chain strategy, it launched several projects designed to reduce the probability that any of these risks would occur. These mitigation actions decreased the estimated risk cost from $22.12 to $12.74 (including the cost of the mitigation activity), giving the team further assurance of the project's feasibility.

Your strategy team should use a process like those previously described to prioritize supply chain risks. This will provide a framework to engage in a group discussion to reach consensus on the subjective issues associated with risk evaluation. After this step, your team will be in a good position to brainstorm ways to mitigate the highest priority risks facing your supply chain.

Mitigate Risks

At this point in your risk management process, you have prioritized the risks faced by your supply chain. In the next step, you need to develop mitigation plans for the top risks. The line organization needs to be deeply involved and own this part of the process. Most often, though, it has time to deal with only the top three to five highest priority risks.

What goes into a risk mitigation plan? Certainly such plans involve some art and some science. The plan should focus on significantly reducing either the probability of a risk occurring and/or the degree of its impact. It could also involve installing an early warning system. Like a serious disease, risk events that are caught early can often be managed successfully. Some elements that companies routinely use in their risk mitigation plans include the following:

- *Insurance*: Firms need to work with insurance providers and create a plan to use insurance to mitigate risk where appropriate, based on an objective cost–benefit analysis (described in more detail later).

- *Best practices approaches*: Companies would be well served to employ one of the best practice models previously described.

- *Inventory*: Some call this "the no-brainer" approach to mitigating risk. It is certainly the most often used, either by design or accident. How much additional inventory results if a source is moved globally without making systemic improvements in the supply chain process? Many of those we talk to say 60 to 75 or more days of supply! Incidentally, some companies try to offset this severe working capital impact by extending payment terms, or using higher payables to offset higher inventories. Of course, this does nothing to address the problem of slow response to customer demand.

- *Expedited shipping*: Some firms accurately realize that "stuff happens" and that they may need to expedite shipments globally in spite of the best-laid plans. Therefore, they prepare thoroughly for that day. In fact, some assume a percentage of the shipments will be expedited or airfreighted when they initially plan for a global source. Knowing this, the proactive supply chain manager may consider investigating different types of insurance coverage. Some policies cover the costs of expedited shipments, depending on circumstances.

- *Import excellence*: Leading companies realize that the better they become at global shipping, the less risk they incur. They strive to achieve import excellence, get the highest C-TPAT certification, and optimize incoterms (international commerce terms, which specify liability and responsibility throughout the global supply chain).

- *Competent partners*: Although it is potentially costly, some companies develop a second domestic source that can be quickly ramped up. They insist on dealing with strong, competent, world-class suppliers, ideally with a "first world" parent. Those who have done this effectively contend that it can take at least two years to develop and certify an excellent source.

- *Financially strong partners*: One major buyer defaulting on a payment could spell disaster for a small to medium-size enterprise. Trade credit insurance, used for many years throughout Europe, is now becoming increasingly popular in the United States. It can help protect domestic and international accounts receivable against unexpected bad debt loss due to insolvency or protracted (slow pay) customers. Companies today are using trade credit insurance as a means to safely and more confidently expand into new markets.

- *Design for globalization*: The simpler the product design and the fewer parts and SKUs involved, the less risk there is in a global supply chain. Leading firms design for globalization. They minimize component parts and SKUs and have rigorous beginning-of-life tollgates and end-of-life processes for their products.

- *Supply chain event management*: An early warning system is crucial if risks are to be identified fast enough to do something about them. Supply chain event management (SCEM) systems put in place criteria that trigger alerts. For example, if a container of critical parts faces a delay at a port, the SCEM system should send an alert to allow the problem to be addressed quickly.

- *Lean/Six Sigma*: When firms combine the principles of Lean and Six Sigma with value-stream mapping, they find a multitude of ways to reduce cycle time and variation in the supply chain by eliminating wasteful activities in the process. Risk diminishes as cycle time and variation decline.

- *Internal functional silo management*: When supply aligns with demand, supply chain risks diminish. Leading companies include supply chain risk management as part of their sales and operations planning process (S&OP) or their integrated business planning (IBP) process.

- *Contracting*: Supply shortages invariably happen. Some firms anticipate the inevitable and work with suppliers to make sure their firms get more than their fair share during serious shortages.

- *Disaster preparation*: The idea is to know whom to call if a natural disaster strikes, such as the American Red Cross, the state office of emergency management, FEMA, etc. In other words: buy the umbrella before it rains.

- *Contingency planning*: Leading companies have documented contingency plans for risks that would have a devastating impact. This would include detailing what would happen if the company lost one of its major suppliers, one of its factories, or one of its DCs.

- *Forward buying (hedging)*: Hedging is a way for a company to minimize or eliminate foreign exchange risk, as well as the risk of commodity price increase. Hedging, though, always comes at a cost.

- *Supplier segmentation*: The idea is to segment suppliers by total financial impact on the firm. This does not necessarily mean total supplier spend. It is clearly possible for a very inexpensive component to shut down a major assembly line. Risk mitigation plans should be developed for the most critical suppliers.

Using Insurance to Mitigate Risk

The least-used strategy in our survey for supply chain risk mitigation was insurance. Unfortunately, insurance does not fall in the comfort zone for supply chain professionals. However, an inexpensive insurance solution can mitigate a wide range of problems, from a port strike to lost cargo. Supply chain professionals we talk to assume that insurance is the purview of other specialists in the corporation. By doing so, they miss a great opportunity to selectively use insurance to mitigate key risks. Insurance provides the financial backstop when a loss occurs, but a company can also leverage the experience and expertise of the broker and the insurance company to prevent losses before they occur. This can be especially useful for the day-to-day challenges previously mentioned. Remember, none of those surveyed uses outside expertise in assessing risk for their supply chain. This must change.

One simple approach supply chain professionals can employ is calculating an "expected loss" for each major supply chain risk:

Expected loss = (cost of loss) × (subjective probability of loss)

Once calculated, expected losses can be compared with the cost of insurance to cover that loss. The expected loss includes the product value, the customer and lost sales impacts, and the expediting costs to recover.

Supply Chain Professionals May Not Understand Insurance Products

Most supply chain professionals have very little expertise with insurance products that can help mitigate supply chain risk. Many we talk to mention that they use carrier liability programs, thinking this is insurance. It is not. Carrier liability programs rarely cover the full value of lost or damaged items. In addition, positive claims settlement outcomes from the carriers can be difficult as the carrier programs typically provide only the bare minimums as required by law.

According to a January 2013 article that appeared in *Inbound Logistics*, "What you do not know about transportation liability can cost you big time . . . plaintiff attorneys are increasingly moving up the supply chain—from carrier to broker, and possibly even shipper—for compensation."[12] This finger-pointing is primarily due to the lack of clarity regarding liability because risk ownership can change throughout the supply chain. It is important to understand who owns what risk, when the risk is owned, and how the risk can be mitigated.

---

[12]"What You Don't Know About Transportation Liability Can Cost You Big Time," *Inbound Logistics Magazine*, January 2013.

Fortunately, insurance programs are available both for parcels and freight/cargo, regardless of the transportation mode or carrier used. These programs go well beyond standard carrier liability limits. They can actually cover high-value, time-sensitive, and temperature-sensitive perishable goods and other hard-to-value items in the event of loss, damage, or delay. Other programs offer consequential loss coverage, provided the shipper can quantify damages.

Summary

Due to its global nature and systemic impact on the firm's financial performance, the supply chain arguably faces more risk than other areas of the company. Risk is a fact of life for any supply chain. In spite of that, the vast majority of companies give this topic much less attention than it deserves.

Supply chain risk cries out for a process to manage it. *We recommend that you implement the three-step risk management process described in this chapter*:

1. *Identify*: Your supply chain strategy team should set aside time to identify the risks facing your supply chain. This should be a free-flowing exchange, with plenty of time set aside. The group should not let themselves feel intimidated or overwhelmed. There are no bad ideas or suggestions. Get everything on the table without any constraints or criticism. A key to the success of this exercise is to identify the right stakeholders. The team should ideally get together in an off-site meeting to recognize the risks facing the firm's global supply chain. This should be done regularly, because risk changes constantly. Once you have solved one risk, another surfaces. Depending on your business, it may be time to advocate for a permanent risk manager who focuses solely on preemptive supply chain risk management and solutions.

2. *Prioritize*: Once you and your team identify the risks facing your supply chain, you should prioritize them to avoid overwhelming the organization. Do not try to solve all the risks facing your supply chain at once. Regardless of a company's size and geographical scope, several methodologies exist to prioritize risks. This study mentions several tools that companies can use, including the FMEA, among others.

3. *Mitigate*: In the final step of a risk management process, mitigation plans need to be developed for the highest priority risks.

The line organization should be deeply involved in and own this part of the process, as should the other stakeholders.

What goes into a risk mitigation plan? Such plans involve some art and some science. The plan should focus on significantly lowering the probability of occurrence and/or the degree of impact, and could include any of the mitigation ideas discussed in the chapter. This includes relying on experts. One of the most surprising findings of this paper was that none of those surveyed used outside expertise in assessing risk. Solutions can come from many different areas, including academia, logistics providers, vendors, insurance companies, and others.

In summary, it should be clear: supply chain professionals cannot afford to delay fortifying their supply chains against disaster. They must install a risk management process now to ensure long-term continuity and resilience for their organizations.

Transition: Managing Risk in the Global Supply Chain to the ABCs of DCs— Distribution-Center Management: A Best Practices Overview

Risk is present in virtually every function of organizational operations. The supply chain is not immune to risk, as was reinforced in Chapter 3.

Risk cannot be eliminated. By its very nature, risk is unpredictable and always evolving. What may have been an inconsequential hazard yesterday could be a devastating disruption today. However, with proper risk management procedures in place, risk can be identified, prioritized, and mitigated. Following this three-step process allows organizations to prepare for the worst. With a risk preparation strategy in place, firms are poised to ride out the very worst of storms and remain competitive throughout.

Distribution centers are just as susceptible to risk as every other supply chain function. Fortunately, there is also a very robust set of best practices distribution centers can follow to not only mitigate risk, but also create efficient, best-in-class facility operations.

Chapter 4 is about the fundamental concepts behind the most effective distribution centers. When it comes to the best companies' distribution-center operations, our research revealed 11 key themes to success. This chapter takes an in-depth look at each of those themes. We define each topic, share examples of real-world application, and provide suggested means for how an organization should go about addressing them.

What good is knowing these 11 key themes, though, without knowing how to assess your distribution center(s) against them? To solve this problem, we have provided a *Distribution-Center Evaluation Tool* at the end of the chapter. This tool asks critical questions as they relate to the 11 themes of distribution-center best practices. Do you have safe processes? Do you have a robust method to manage returns? By the time you have finished the chapter and completed the evaluation, you will have a much better idea of where your distribution center(s) stand.

Game-changing supply chains all possess proper distribution-center management. Effectively managing these distribution centers can be the difference between average and world-class operations. As the trends identified in the first chapter (and its addendum) continue to evolve, effective distribution-center management will require new management techniques that address the themes identified in this chapter.

The ABCs of DCs is meant to help you change your distribution-center game. We hope you find the topics contained within, as well as the assessment at the end of the chapter, to be of use.

4

The ABCs of DCs—Distribution-Center Management: A Best Practices Overview

The Fifth in the Game Changers Series of University of Tennessee Supply Chain Management White Papers

J. Paul Dittmann, Ph.D., Lloyd Rinehart, Ph.D., Ted Stank, Ph.D., Chad Autry, Ph.D., Mike Burnette

Introduction

Logistics professionals who operate distribution centers (DCs) have a tough job. Management constantly challenges them to cut cost, which means doing more with less. While focused on that, they need to make sure that customer responsiveness does not suffer and even improves. Clearly, this calls for a highly advanced management skill set.

This chapter draws on data from three sources. The first is information gleaned from the supply chain audits done at the University of Tennessee. The second is from third-party logistics professionals who manage extremely large warehouse networks for very demanding clients. Finally, the third set of inputs was drawn from industry supply chain professionals with extensive warehouse management responsibilities. The participating companies included prominent manufacturers, 3PLs, and some of the largest retailers in the world.

Eleven themes emerged from this research and are covered in the chapter:

1. Receiving, put away, and returns

2. Picking, order fulfillment, and shipping
3. Lean warehousing
4. Cross-docking
5. Metrics and planning
6. Warehouse information systems
7. Warehouse layout and space optimization
8. Warehouse network optimization
9. Safety and security
10. People
11. Sustainability

For each theme, we include a short discussion of best practices for supply chain professionals to consider as they develop a DC management strategic plan. DCs come in many shapes and sizes, from case picking to individual item; highly automated to mainly manual; small in square footage to just plain huge. To cover this scope in one chapter is daunting, but the best practices described below are intended to apply to the broadest possible range of DC types.

Receiving, Put Away, and Returns

All warehouses need an efficient process to receive and put away, or cross-dock goods. The receiving and put-away processes critically affect overall warehouse efficiency. Cross-docking is discussed in its own section later.

Receiving

At the most basic level, vendors that ship into DCs need to be reliable. Suppliers must be certified so that quality and accuracy are assured. Inspection and count verification required at the DC should be minimal. Suppliers should also have the capability of sending advanced shipment notices (ASNs) before their shipments arrive.

World-class receiving is highly facilitated by the use of ASNs. An ASN notifies the DC of a pending delivery and is usually sent in an electronic data interchange transmission. Suppliers use ASNs to list the contents of a shipment as well as additional detailed information describing the shipment's composition and configuration. By receiving the ASN before delivery, receiving cost can be reduced and accuracy improved. ASNs make labor planning much easier since DCs know what will be hitting their docks before it gets there. It should be noted that many manufacturing companies receive product from their own factories to their own DCs. In those cases, the internal systems can be linked, accomplishing the same thing as an ASN.

ASNs eliminate most data entry, and data entry errors, at the time of receiving. DCs' receiving operations can do quick scans of barcodes on shipping labels and electronically match them to the ASN information. Generally, an ASN provides a list of all the barcoded identification numbers of the shipment contents. This improves inventory accuracy and greatly reduces receiving costs. Cost-reduction estimates are in the 40% to 50% range.

In addition, the advanced shipping notice can be used to pay suppliers directly for goods received by inputting the notice into the company's enterprise resource planning system, which includes the accounts payable system. Any discrepancies can be quickly transmitted to the supplier and corrected, so payment can be executed in a timely manner.

The use of advance shipment notices is clearly a best practice. ASNs have been around for decades and one would think their use would be nearly ubiquitous by now. Unfortunately, that is far from true. Progress is being made, but shockingly some of the largest and most prominent companies still have not fully implemented ASN technology in their finished goods DCs. One major retailer told us that 30% of their inbound shipments had an ASN, with the major constraint being the capability of their suppliers. Another major retailer said that their suppliers could provide ASNs, but their internal systems could not handle them. ASN technology is far more common in factory operations for the receipt of raw material and less common in finished goods DCs.

Put Away

A modern warehouse management system (WMS) should provide exact direction for put away. Goods should be placed in the best locations to facilitate picking (see profiling/slotting discussion later), and they should be placed in warehouse locations to minimize the travel distance and time of DC personnel.

Best practice DCs put away product quickly. Companies should measure "dock to stock" time to help facilitate this process. Slow put away negatively affects space, causes congestion, increases transaction errors, and makes product more susceptible to damage. The most efficient DCs move product directly from receipt to the final location.

Direct put away programs require a good WMS that can assign locations from an advanced shipping notice or upon receipt to the dock.

Best practice companies also use integrated engineering standards in their WMS and pick locations and replenishment areas so that an optimal put away route from receiving to storage areas can be selected. Many WMS programs also support task interleaving so that put-away and picking operations can be performed in tandem to greatly reduce nonproductive travel time.

Returns

Returned products need to be received as well. With the pan-industry race to make returns easier and more customer friendly, as well as the increase in Internet sales, DCs can expect returns to increase substantially. Of course, many supply chain professionals believe that the best practice is to eliminate returns—or at least returns back to the DC. This is generally done by offering a return allowance in order to incentivize the retailer to deal with returns. Alternatively, some firms farm out handling returns to a company that specializes in it, such as Inmar. Other companies feel that the customer relations risk requires returns to be handled in-house. Ironically, in many such cases, returns are often treated as an afterthought and managed in a highly disorganized way. When it comes to managing returns, many companies leave a lot of money on the table. A little attention here can pay big dividends. A recommended process for handling returns is described in Figure 4.1 below.

Picking/Order Fulfillment/Shipping

Aggressive DC productivity goals cannot be met without an efficient process for picking or order fulfillment. In fact, picking often consumes the lion's share of DC labor. Based on our statistics, warehouse labor is consumed in the manner illustrated in Figure 4.2.

To optimize warehouse efficiency, the DC needs to focus on and minimize three things: travel distance, touches, and paper. The best picking systems have low rates of each. Of course reducing travel distance is critical, and creativity is helpful. For example, one company told us that it installed forty-five degree aisles in their DC and saved a major amount of time by traveling the hypotenuse.

In our discussions with supply chain professionals, we saw a wide variety of ways companies efficiently fill orders. Of course many small

A recommended process for managing returns

- **Identify** root causes of returns.

- **Measure** the full cost of returns, which most companies grossly underestimate.

- **Review** product design, packing, and consumer instructions.

- **Manage** better customer education and expectations on the front end of the purchase.

- **Segment** returns with a different approach for each category.

- **Develop** an operations plan to minimize the processing cost.

- **Put in place** a liquidation plan to maximize asset recovery; a decision tree framework is often useful.

Figure 4.1 Key elements of returns management

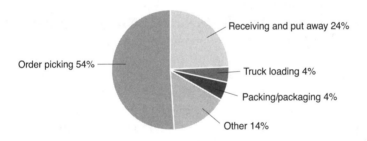

Figure 4.2 Order picking drives labor cost

DCs have not abandoned paper pick lists. But whether done with paper or with the technologies described below, a key to efficient picking is proper profiling or slotting.

Warehouse space optimization and layout closely relate to profiling, and they are also fundamental to an efficient picking operation. We discuss this theme in its own section later in the chapter.

Profiling/Slotting (Zoning or ABC Zoning)

It goes by various names, but regardless of the moniker, Profiling/Slotting is a prerequisite to efficient fulfillment operations. Many DCs dedicate full-time personnel to the task of profiling or slotting. Their job is to ensure that high velocity SKUs are placed in convenient, easy to reach areas to minimize pick times. They do this by studying the velocity of each SKU stored in the DC, often with the assistance of their WMSs. Studies have shown that travel time is three times more impactful than search and selection time, although both are important and both are facilitated by a good profiling system. Warehouse layout and picking are closely connected; in fact, a primary goal of warehouse layout is to facilitate picking.

Profilers know that it costs more to pick vertically above the floor than horizontally on the floor. The fastest moving items are thus assigned to floor level. Hot zones are set up for fast moving SKUs to increase picking productivity. High cube items are often assigned to racks, while low cube, smaller items are assigned to bins or shelves.

We saw a good example of a profiling operation at an office supplies retailer's DC. They dedicated three people to ensuring that the right SKUs are in the right places. They use a feature in their WMS to assist in this effort. Others use special slotting or profiling software. Such software can be configured to physically separate error-prone items where pickers tend to mix items if slotted too close together. A cosmetics manufacturer has five rows of items in its pick area. It puts the highest volume lipstick in row three at eye level, but separates similar colors to avoid pick errors. An appliance manufacturer calls their version of profiling "ABC banding," which ensures that they put the highest volume appliance SKUs closest to the outbound truck docks.

As noted, profiling should be done on a dynamic basis with the aid of the WMS. There should be no permanent locations, given the frequent changes. One caveat: fast moving items should be easy and fast to pick. But they may have to be spread out to minimize congestion problems.

Picking Systems

One of our recent University of Tennessee surveys found that firms used the following types of picking systems:

- Radio frequency vehicle mount—80%
- Paper pick list—40%
- Pick to light—38%
- Carousels/conveyor systems—25%
- Voice picking—5%
- A frame—5%
- Automated storage and retrieval systems—5%
- KIVA—1%
- Other—22%

Voice Picking

Some call this the greatest gain in picking in the past twenty-five years, while others are much more subdued in their enthusiasm. With voice picking, operators wear a small portable computer with a headset incorporating a microphone. They receive verbal commands via the headset and confirm actions through the microphone to the WMS.

When it comes to accuracy, voice picking wins, because the operator reads a check digit back to the computer to confirm the right pick was made. Error rates of less than 1-per-1,000 picks have been reported. Other picking technologies are accurate, but if you go from 99% accuracy with one method to 99.9% accurate with voice, you avoid 9,000 errors per million picks. This can clearly be significant for some products and customers.

Voice systems can speed up the learning curve for new warehouse employees as well. One firm told us that the time between on-boarding and full efficiency went from five weeks to five days when they converted to a voice system.

Installing a voice-picking system requires a significant investment, perhaps $5,000 per unit. So voice picking is most appropriate in a larger facility with a significant number of people. To generate a high enough ROI on those sunk costs, a rule of thumb would be forty to fifty workers using the system.

Voice picking is hands-free and therefore also contributes to safety. It can employ multiple languages within the same warehouse, which is beneficial for example to US operations in areas with a high Spanish language population. Having native language capability enhances operator productivity. The voice speed in the headset can also be set very fast as operators become accustomed to it. Although high-speed voice seems incomprehensible to rookies, highly experienced operators quickly adjust to it.

Pick-to-Light

With pick-to-light, stock locations have light nodes connected to the main computer system. These light up indicating which units to pick. When the associate completes the pick, he or she presses a button next to the light to confirm it.

If applied in the right setting, pick-to-light systems are faster than (but not as accurate as) voice. For small items, especially in small slots, someone could touch the light but reach into the wrong slot. This can happen an estimated three times in every 1,000 picks—at least three times higher than voice picking. Weight check systems at the end of the pick can help catch such errors.

Of course, speed depends on the right application. Pick to light works best for small, relatively fast-moving items where a picker can simultaneously see all items to be picked for an order, rather than be given instructions sequentially. It is not as useful for case picking or picking a wide range of larger product sizes/weights. Pick-to-light systems can require expensive conveyors, challenging ROI.

Radio Frequency

In wireless radio frequency picking environments, the associate generally has a radio frequency gun that can be holstered. Radio frequency/scanning is highly accurate but not perfect. Sometimes an operator can take his or her eyes off the picking location when the scanner is being returned to the holster. Items and pallets are scanned, and that is matched to a scanned location to enhance accuracy. Voice and pick-to-light systems are more productive and more accurate than radio frequency. Radio frequency pickers spend an estimated 15% of their time manipulating the radio frequency gun.

Automated Picking (A-frame or Automated Storage and Retrieval Systems)

A-frame picking systems have bins or totes that flow on a conveyor through a stocked A-frame. Based on the bar code, the right units fall into the tote as it passes under the A-frame. A-frames are generally used for small items.

Automated storage and retrieval systems can be several stories tall and totally automated. A robotic device moves through the high-rise warehouse picking the appropriate units from the warehousing racks.

Automated systems require a major capital investment. So without a long contract in place, 3PLs find these systems to be problematic.

Goods-to-Person Picking (KIVA)

With KIVA, a wholly owned subsidiary of Amazon, robots bring the product to the person assembling the order. A rule of thumb is that at least 50% of picking time is operator travel time, often much more. These systems eliminate operator travel time. A number of companies other than KIVA now offer goods-to-person robotic systems. These are expensive. An operation would have to be a large, high volume facility often involved in each picking (picking individual units, not cartons) for this technology to pay off.

Robotics and AGVs

Robotics are becoming more feasible as technology advances. They can do basic picking and other operations. For example, hourly associates can easily program Baxter robots from ReThink Robotics. They are relatively low cost at about $25,000. And, if they can replace a full-time employee, the ROI should be quick.

Automated guided vehicles (AGVs) are another form of robotics. In the original models, the units followed a wire embedded in the floor. Newer AGVs can be laser guided. Mirrors and laser beams continuously tell the vehicle where it is. Some newer AGVs accomplish this with location bar codes around the DC, along with a camera/optical recognition system.

Manual/Paper-Based Systems

These are the least accurate and the least productive systems. But a paper-based system may make perfect sense for many small operations, and there are thousands of very small warehouse operations employing ten people or fewer.

Postponement Operations

Picking and order fulfillment will be heavily impacted by the trend toward postponement. More and more orders will be kitted, packaged, given unique identifiers like branding and literature, or lightly assembled. These capabilities are driven by the vision of a low inventory, flow through environment, and they will become increasingly common in the future e-fulfillment world. In postponement and other DC activities, packaging must be optimized to reduce the waste, weight, and cube.

How to Select the Right Picking System

When deciding which picking technology to employ, several factors come into play. Three of the most important are operating cost, accuracy, and capital investment. Safety should also be considered along with product damage potential. When calculating the ROI of various options, it is important to put a number on hard-to-quantify items like safety or picking accuracy. Otherwise, they tend to become nonfactors in the decision when in fact they may be among the most important considerations.

To narrow options prior to doing an in-depth ROI analysis on the most viable alternatives, create a matrix like Table 4.1 and rate each cell:

Table 4.1 Radio Frequency is the Most Popular Form of Order Picking

How to Select the Right Picking System

When deciding which picking technology to employ, several factors come into play. Three of the most important are operating cost, accuracy, and capital investment. Safety should also be considered along with product damage potential. When calculating the ROI of various options, it is important to put a number on hard-to-quantify items like safety or picking accuracy. Otherwise, they tend to become nonfactors in the decision when in fact they may be among the most important considerations.

To narrow options prior to doing an in-depth ROI analysis on the most viable alternatives, create a matrix like the following and rate each cell:

Decision criteria	Decision criteria weighting	Picking System A	Picking System B	Picking System C
Operating cost/ productivity	Weight	Ranking	Ranking	Ranking
Capital investment	Weight	Ranking	Ranking	Ranking
Integration with other systems	Weight	Ranking	Ranking	Ranking
Accuracy	Weight	Ranking	Ranking	Ranking
Safety	Weight	Ranking	Ranking	Ranking
Damage	Weight	Ranking	Ranking	Ranking

Note that the table above has a column labeled "Decision Criteria Weighting." This forces the selection team to make hard decisions regarding their most important needs in a picking system.

Once the options are narrowed to perhaps two or three picking systems using the ranking process above, it's time to gather hard data from vendors, probably through an RFP process. After the cost/investment and benefits are accurately quantified, an accurate ROI for each option can be calculated and the selection made.

Note that Table 4.1 has a column labeled "Decision Criteria Weighting." This forces the selection team to make hard decisions regarding their most important needs in a picking system.

Once the options are narrowed to perhaps two or three picking systems using the ranking process above, it is time to gather hard data from vendors, probably through an RFP process. After the cost/investment and benefits are accurately quantified, an accurate ROI for each option can be calculated and the selection made.

Omni-Channel Picking

Omni-channel fulfillment—from Internet and mobile phones sources—is becoming increasingly common for manufacturers and retailers. The rise of the omni-channel means that DCs can no longer batch all of their orders into large waves that must be completely processed before taking on the next wave of orders. Internet orders are filled in a waveless picking process. They have to be filled very quickly, perhaps within a couple of hours, and often result in difficult piece picking, exacerbated by seasonal spikes. It is further burdened by a large number of SKUs and creates the need for increased inventory control, flexibility, and fast, accurate fulfillment.

Instead of one large order to pick for a store or another warehouse, DCs will have to ship hundreds of small orders to individuals all over the country. Warehouse operators will have to become increasingly flexible in their pick, pack, and ship operations. Speed, efficiency, and accuracy will be required as never before. No doubt, creative automation approaches will emerge. More and more large retailers are filling Internet orders out of retail store locations as well as their DCs. Software exists to determine the optimal ships from a location. The customer can get home delivery for an additional charge, or pick up the order at the retail store location.

Some retailers are partnering with services like Google Express to fulfill their Internet orders. Google Shopping Express is a same-day shopping service that was launched on a free, trial basis in San Francisco and Silicon Valley in spring 2013 and publicly in September that year. In spring 2014 it was expanded to New York and Los Angeles, and in fall 2014 to Chicago, Boston, and Washington, DC. Google does not have any physical infrastructure, and instead sends shoppers into existing stores.[1]

---

[1]"Google Express," *Google*, https://www.google.com/shopping/express (retrieved May 2015)

Lean Warehousing

Lean concepts originated in the 1950s in Japan and developed and matured in Toyota factories over several decades. The Lean philosophy first reached the shores of the United States in the 1980s with a few pioneering companies like John Deere and Harley Davidson. Then it exploded into US manufacturing in the 1990s. From there, Lean concepts escaped the confines of the factory. The Lean philosophy is now being used throughout the supply chain, especially in warehouse operations.

Many firms, both retailers and manufacturers, are now aggressively rolling out Lean in their warehouse operations. In the process they reduce cycle times, speed up customer responsiveness, and reduce waste throughout their operations. Huge paybacks are being seen with Lean implementations in one warehouse operation after another. It is safe to say that, if a warehouse operation is not implementing Lean, it is falling behind its competition.

Many companies have their own versions of Lean that closely follow its fundamental principles, such as the Kenco Operating System or the Honeywell Operating System. If the 3PL is running the DC operation, they will need to take the first steps toward implementing Lean in the facility. Some examples of Lean concepts commonly used in DCs are discussed below.

Keep It Simple

Do not over complicate this. The Lean journey should focus on straightforward activities that everyone can see and understand. This means concepts like 5S, total employee involvement, standard work, visual management, and management walkabouts. This foundation can then lead to more sophisticated concepts such as value-stream mapping. Lean is a lot about total employee involvement in a continuous improvement effort, so DCs should not rush to implement the Lean tools. Sometimes the tools take on a life of their own, causing the operation to lose sight of the simple goal of waste elimination. With that said, some of the tools and concepts of Lean are described below.

5S

5S refers to a methodology and mindset that ensures that all areas are neat and clean, with everything in its place. Although it varies, the five S's usually stand for "sort, set in order, shine, standardize, and sustain." A number of warehouse operations add a sixth "S" for safety. Few would argue with that. In fact, in most warehouse operations, the warehouse manager is quick to say that safety is the number one priority.

Total Employee Involvement

The real secret of Lean is that it creates a learning organization that focuses on continuous improvement. Lean takes a full commitment from the top and full involvement from the bottom. One company chose several "lean champions" from among the hourly associates in the DC and asked them to gather improvement ideas from their peers. Those Lean champions also led "5 Why" sessions. In these sessions, the problem is identified, and then the group asks "why" five times to reach its root cause. One firm has a simple document called a SPIN (simplified process improvement needed) form available for hourly associates to submit their suggestions. This firm said that this process starts slowly but really takes off when people see their ideas being implemented.

Standard Work and Making Problems Visible

Some Lean advocates say that there can be no kaizen without standardization. All work needs to be rigorously documented. That ensures continuity of improvements and makes deviations more visible. The three-step process of "standardize, measure, and improve" has standard work at its root.

Some have used the Maynard Operation Sequence Technique system with success. The MOST system is a predetermined motion time system that is used primarily in industrial settings to set the standard time in which a worker should perform a task.

Management Walkabouts

Some years ago, managing by wandering around was all the rage—and for good reason: it worked! Managers need to get up, get out and about, and talk to people face-to-face. It is the only way to understand the problems and the mood of the organization. It is important to do this with a set of good questions, such as "Do you see anything we could do better?" or "Is your area ahead or behind schedule?"

Millennials especially may need encouragement to give up electronics and talk directly to humans.

Visual Management

Lean warehouse operations are visual operations, designed to promote employee involvement. As one warehouse professional asked, "How can you expect your team to perform if they do not know the score?" Lean operations visibly place charts and graphs everywhere. Andon lights clearly show the status of mechanical operations. One company mounted a large four-foot-by-eight-foot poster displaying a fishbone diagram that focused on an inventory accuracy problem existing in the warehouse. On the horizontal spine of the diagram was shown the problem (i.e., inventory accuracy). On the diagonal bones of the diagram were listed general areas that could be the source of the problem (like systems or human error). All associates were invited to use sticky notes to comment on various aspects of the problem.

Total Productive Maintenance

With total productive maintenance, associates who work with equipment are also responsible for routine maintenance. This not only promotes employee involvement and ownership but also reduces unexpected downtime. As one associate said, "Who knows more about how your car sounds: you or the mechanic in the shop? It is the same with the equipment we use every day."

It is critical that conveyor systems and automation do not break down unexpectedly. Therefore, maintenance technicians should focus on predictive and preventative maintenance, rather than routine maintenance. Routine maintenance should be left to the equipment operators. Maintenance technicians should plan and execute preventative maintenance activities during planned downtime. And they should be using advanced techniques such as vibration analysis and infrared analysis to detect problem areas, as well as hot spots that could predict failure.

Eliminate Before You Automate

Automation is critical to large, efficient warehouse operations, but it can be overdone. One company had an empty automated storage and retrieval system, calling it the "epitome of a monument to waste." They had found a way to flow product through the warehouse without putting it in, and pulling it out of, the automated storage and retrieval systems. Another company eliminated the need for an automated guided vehicle by simply moving the two material points next to each other. An "eliminate before you automate" mindset avoids hardwiring waste into an operation.

Value Stream Mapping

Value stream mapping helps identify value-added versus non-value-added activities in a process. It is used as a vehicle to eliminate waste. It requires visually mapping a process and then asking a group of employees/associates to identify opportunities for improvement as well as non-value-added activities to eliminate. Value stream mapping can ensure that there is quality at the source and that there are no rework loops. It can be applied to any process and almost never fails to identify savings of at least 20% or more (according to the experts we polled). It is critical to include in the value stream the supply chain steps both prior to the warehouse wall and after the warehouse wall. The biggest losses can be in the handoff between supply chain activities.

Much like the fishbone diagram example above, one warehouse created a four-foot-by-eight-foot map of their receiving process. Employees were encouraged to place sticky notes with their ideas for improvement as a lead-in to the value stream mapping exercise.

Office Operations

It should be noted that some DCs have applied Lean concepts in their office operations, including standard work, visual management, and value stream mapping. This could encompass anything from office proximity and printer placement to checklists and work in progress boards.

Six Sigma

Six Sigma was never part of the Toyota Production System or Lean. It is instead a separate set of tools and philosophies that nicely complement Lean. In fact, many companies call their program Lean/Sigma or something similar. In many companies Six Sigma simply means a measure of quality that strives for near perfection. It is a disciplined, data-driven approach and methodology for eliminating defects and variation in any process. A product of Six Sigma quality has 3.4 defects/variations per million or is consistent 99.9997% of the time.

The Six Sigma DMAIC system improvement process includes five steps for processes falling below specification: define, measure, analyze, improve, and control. Six Sigma has a similar methodology for projects in development called DMADV (define, measure, analyze, design, and verify). Six Sigma black belts learn to apply analytical and statistical techniques to business problems. In many companies, they have established a track record of multi-million dollar savings.

Cross-Docking

Cross-docking is the process of receiving product and shipping the product out the same day without putting it into storage. Since picking and put away consume the lion's share of cost in a typical warehouse operation, productivity skyrockets if those two activities can be eliminated. The cross-docking practice also frees up warehouse space and speeds service to the customer. Some high volume warehouse operations are designed to cross-dock automatically. One large retailer receives cartons from suppliers and moves them from the inbound truck into a high-speed conveyor system. Many of those cartons continue to flow all the way down to a truck waiting to go to the retail store. There is no picking or put away for such items.

However, most warehouse operations do not have the luxury of high-volume case flow. If they do cross-dock, they have to do it the old fashioned way: they schedule inbound delivery from their suppliers to coincide with outbound delivery. That is far easier said than done, and it is far easier if central control of orders and inventory exists. One retailer told us that they cross-dock about 40% of their goods using such a centralized system. The corporate group has visibility to each inbound purchase order and can dynamically allocate it to an outbound trailer when it arrives. At another extreme, some pure cross-dock locations are set up as staging areas and are not meant to carry inventory. These might be used to receive a full load from a DC into a metro area and break it up at the cross dock into local delivery loads.

In general, a very small percentage of product is being cross-docked today, even though many warehouse managers—up to two-thirds—report that they opportunistically cross-dock to some extent. DCs do not want to tie up trailers, and especially tie up drivers, waiting for a cross-dock opportunity. Many companies turn to their 3PLs to manage the cross-docking operations. Others use a mixing center. A sophisticated WMS can help, but the most advanced technology is useless without a lot of front-end work.

To execute a feasible cross-docking operation, companies need to lay the foundation with:

1. *Reasonably predictable demand*: without knowing demand, sizing and staffing the operation are difficult.
2. *Appropriate products*: ideally, the product involved would have a single, efficient method of handling.
3. *Reliable, efficient suppliers*: suppliers must deliver complete orders, on time, within narrow time windows. To do this, the customer must give their suppliers a stable order pattern and high volume, single SKU loads (ideally).
4. *Expert service providers*: fast, reliable, consistent service is a must.
5. *Advanced systems*: cross-docking is facilitated by sophisticated systems, including automated shipment notices, warehouse and yard management systems, a cross-dock management system, and track and trace capabilities.
6. *Facility design and layout*: inbound and outbound doors and facilities should be designed to maximize efficiency.
7. *Visibility*: easy access to current day inbound and outbound allows a DC to opportunistically match them when possible.

This daunting list requires a multi-year journey to realize the full benefits of cross docking. That said, companies can (and should) get started by opportunistically cross-docking without having all elements perfectly in place. As firms look to make the next big advance in DC productivity, cross-docking merits a very close look.

Metrics and Planning—Developing a DC Management Strategic Plan

The DC strategic plan needs to be consistent with the overarching supply chain strategy. Unfortunately, our research shows that only 16% of companies have a documented, multi-year strategy for their supply chain. Paul Dittmann's book, *Supply Chain Transformation:*

Building and Executing an Integrated Supply Chain Strategy (McGraw Hill, 2013), sets out a nine-step process for creating such a plan. Those nine steps, which have been slightly modified for DC management, are listed below:

1. Start with your customers; understand their needs
2. Assess your internal warehouse capabilities versus best-in-class
3. Evaluate the DC management game changers (i.e., the topics in this chapter)
4. Analyze your competition's warehousing capabilities
5. Survey DC management technology
6. Manage risk by taking the time to identify, prioritize, and mitigate the high-priority risks facing your warehousing operations (see Chapter 3)
7. Determine new DC management capabilities, and develop a multi-year project plan to implement them
8. Evaluate the organization, people, and metrics
9. Develop the business case and generate buy-in

Metrics

Step eight in the strategic planning process includes choosing the best metrics that will motivate a successful warehousing operation. In our surveys, supply chain professionals from a broad range of companies used a scale of 1 to 10 (10 being the most important and 1 the least important), to rank the following supply chain issues, and the result of that is shown in Table 4.2 below:

Table 4.2 SC Professionals Rank the Warehouse Critical Metrics

Implementing the right metrics and setting the right goals	8.15
Establishing collaborative relationships with suppliers and customers	7.91
Advances in supply chain visibility	7.80
Professional development, training, education	6.71
Helping with revenue generation	6.62
Managing the global supply chain	6.55
Effectively using technology	5.21

Make Sure Metrics Have a Logical Framework

The quest for the best metrics is clearly foremost in the minds of supply chain professionals. Key performance indicators need to be linked in a logical framework to the goals of the company. Otherwise, they are simply a laundry list of items with no apparent logic. For example, if the prime goal of the firm is to drive shareholder value, then a metrics framework should be established so that the organization clearly sees how every measure flows into and drives shareholder value.

In developing new metrics to support the strategy, a set of criteria must be in place to avoid poorly designed or seriously flawed metrics. For example, one firm we worked with defined a set of criteria to design the new metrics for its supply chain strategy. The metrics had to reasonably satisfy the following criteria in order to be part of the key performance indicator framework:

- Stable and accurate data with few large, random, or unexplainable swings
- Understandable to everyone with a line of sight so that key personnel can see how their actions influence the metric
- Designed so that they cannot be easily manipulated or gamed
- Capable of drill-down analysis so that root causes of changes are apparent
- Clear cause and effect drivers
- Easily accessible for relevant parties, and available in clear reports, developed and published with clear explanations

These criteria served as a rigorous screen before a new metric was adopted and ensured a small number of very high impact key performance indicators.

Goal Setting and the Importance of Benchmarking

Selecting the right metrics and defining the associated responsibilities is important. Establishing goals is an entirely different matter. Too many companies only use internal comparisons (this year versus last year, for instance), and feel good about achieving an internal goal. Leading companies benchmark best-in-class performance and then set goals accordingly.

The Best Warehouse Metrics

We asked a wide range of DC management professionals their opinion of the best metrics for warehouse management. The responses centered on five areas: safety, customer service (on time delivery, order cycle time, order accuracy, and damage), cost/productivity, asset management, and people development/morale.

Safety

Safety, covered later as a separate theme, should be the number one objective of any operation, and warehouses are no exception. Safety is a mindset reinforced in many ways.

Safety is often measured as the number of incidents or recordables, lost time accidents, times since the last accident, or near misses. Near misses are a very valuable metric giving a leading indicator of recordables or lost time accidents. To capture near misses, a firm has to rely on the commitment of its management team and hourly associates to record a near miss in writing and submit it to the system. Safety metrics should be highly visible to the entire workforce.

One company included safety metrics on the balanced scorecard of every employee in the supply chain organization, including planning and inventory management—not just the logistics operations people. This heightened the awareness of the entire supply chain organization to safety issues.

Customer Service

Some organizations group customer service as part of their quality metrics because failure to serve a customer's needs is clearly a quality defect. Therefore, customer service key performance indicators are ideally expressed in defects per million. Ninety-nine percent good seems acceptable until one realizes it represents 10,000 failures per million. Customer service metrics tend to be focused on:

- Order accuracy and pick accuracy
- Damage
- Order fulfillment time vs. standard
- Incorrect paper work, including the invoice
- On-time shipment, or on-time delivery to a customer's service window

The best practice is for customer service to be measured and reported by the customer.

Cost

Cost metrics tend to be dominated by productivity measures, since labor generally represents most of the facility's operating cost. It is important to break cost down into its main components: receiving, put away, picking, shipping, maintenance, and management. Typical key performance indicators that measure cost productivity are units or cases per man-hour, picks per man-hour, cost per pick, cost per carton, throughput rates, etc. One firm said that they measure cost per "blended unit," where a blended unit represented one item picked. It could be a case, or it could an individual item. Another firm said that they moved to a "cost as a percentage of sales." They did that in order to make sure the logistics operation cared as much about sales and revenue generation as it did about cost. Another best practice is to use a weighted average cartons/units picked per hour, because all cartons/units were not created equal.

Time measures can also be quite powerful, such as time from order receipt to fulfillment and then time to shipment. Also important is time from inbound arrival to unload and time to put away.

One company found that productivity metrics alone drove an excessive investment in automation. Management remedied that by including the depreciation cost of the automation in its total cost analysis.

Capturing and tracking cost metrics beyond direct labor is also important. This includes items like indirect labor, contractors, overtime, and health care.

Some companies compare DCs in an ordered ranking by key metric. While this is somewhat unfair since nothing is ever apples-to-apples, it is still motivational. No one wants to be at the bottom of any such list, and anyone at the top wants to stay there.

Since DCs are responsible for loading trailers, they assume the burden of making sure that the equipment is as fully utilized as possible. One retailer who rigorously manages cube utilization gets a 3,000 cube in the average 53-foot trailer (a trailer has 3,800 cubic feet of space). This very positively impacts transportation cost.

Asset Management

Assets consist of physical capital and working capital. Physical capital includes warehouse space as well as the equipment, automation, and systems used to operate the warehouse. Working capital consists mainly of inventory, plus receivables and minus payables. Maintaining accurate inventory is a critical responsibility of warehouse operations. Leading companies not only measure if the right inventory is in the building, but also if it is in the correct place in the building. On-time delivery triggers the invoicing/payment process, and therefore initiates the receivables process, receivables being a critical component of working capital. Procurement, a supply chain function, manages payment terms to suppliers, and therefore strongly influences accounts payable. Therefore, supply chain has a major impact on all three buckets of working capital: inventory, receivables, and payables.

People Development, Management, and Morale

The concept of a balanced scorecard means that qualitative metrics are also critically important. These often come into play in the people development and management area, which we cover in the People section in this chapter. Some examples of important metrics to consider are:

- Evaluations completed
- Professional development plans in place
- Education and training completed
- Turnover and reasons for it; one firm estimated a $9,000 cost for each hourly associate replaced, not counting the learning curve effects; turnover also is a good surrogate measure of morale.
- High potential employee development plans completed

Warehouse Information Systems

A modern WMS is critical to the efficient management of any medium-to-large-sized warehouse. One large retailer even installed a WMS in their retail store's backrooms, especially after the supply chain organization assumed responsibility for moving product to the retail shelf. While this may sound like gross oversimplification, WMSs essentially improve productivity by minimizing movement in the warehouse. The idea is to reduce operator and equipment movement as much as possible to minimize the cost and number of people required to run the warehouse. This arguably leads to an even more important result: the speed-up of customer order fulfillment times.

Warehouse Management System (WMS) Functionality

To choose the correct WMS, the needed functionality must be identified first. Some of the most common capabilities of WMSs and common bolt-ons are:

Goods receipts

The best practice is to match the advanced shipment notification to the bar codes, often using a handheld device. The system then indicates the best location to store the goods and should also trigger the vendor payment cycle. In FDA-compliant facilities, during the goods receipt process there is an additional step of quality control. Inbound loads are inspected to ensure that they are not damaged, have the proper labeling, are released from the customer for storage, and meet expiration date business rules.

Put away

The system optimally routes any lift truck vehicles, as well as operators on foot, to minimize distance traveled. The system helps consolidate locations during the put-away process and optimizes space utilization by ensuring SKUs are placed in locations that need to be topped off instead of choosing a new empty location. It optimally places the SKU in a location that supports velocity movement and considers any promotional or substitution business rules in deciding where to place SKUs in the DC.

Profiling or slotting

The WMS can analyze SKU movement velocity and recommend where to place SKUs in a warehouse to minimize picking requirements and congestion.

Picking

The WMS can ensure that movement is minimized in picking an order. It interfaces with, and drives, any picking technologies. The WMS allows for multiple orders to be picked simultaneously to minimize operator travel distance. This practice is referred to as "task interleaving." Task interleaving can combine put-away tasks with picking tasks. For example, a driver has a put away of SKU A on Aisle X and there is an order for SKU B that is also on Aisle X. Task interleaving would immediately give the picking task to the driver as soon as he confirmed the put away, allowing him to stay in the same area.

Shipping

A final check matching data in the WMS should be made at the dock to ensure 100% accuracy in the shipment. Some WMSs, or associated bolt-ons, have a capability to optimize trailer loading or facilitate loading, particularly in a multistop environment.

Real-time visibility into inventory and orders

The WMS interfaces with the enterprise resource planning transactional system to provide real time visibility of inventory and orders.

Labor management

Labor management systems are often embedded in, or bolted onto, the WMS. Labor management software reports on the performance of individual associates against discrete standards or goal times for tasks in the DC. A labor management system can analyze historical data for an accurate estimate of warehouse throughput. Then it can schedule the right balance of overtime, as well as regular and temporary labor, required for shifting demand patterns.

Yard and dock management

The WMS and yard management system direct inbound trailers to a particular dock door to minimize receiving and put-away time. Outbound trailers are called to docks as they are needed, based on preestablished priorities. Trailers are checked into and out of the yard, and sealing and unsealing of equipment is monitored. All yard stock should be visible. Some yard management systems utilize active radio frequency ID tags to quickly identify what is inside a truck and where the trailer or container is located. The WMS/yard management system can allow only authorized personnel to use gas pumps, and then record the amount of fuel pumped. It can interface with electronic seals and send an alert if a seal has been broken.

Cross-docking

With a cross-docking capability, inbound loads are matched with outbound deliveries and cross-dock opportunities are dynamically and opportunistically detected.

Inventory cycle counting

Routine cycle counting is managed to achieve excellence in inventory accuracy. Warehouse personnel are directed to count SKUs on a rotating basis throughout the year. Sophisticated cycle counting systems have different rules by SKU family (i.e., fast movers are counted on a more frequent basis).

Integrated with enterprise resource planning system

Integration provides a seamless transfer of order data to the WMS, and a transmission of shipment and inventory data to the enterprise resource planning system.

Reverse logistics

The WMS can also manage returns. If left to ad hoc manual processes, returns can create major inefficiencies in warehouse operations.

Small parcel management

This is typically a bolt-on to the WMS and can choose the best method for shipping small parcels, Fed Ex, UPS, and US Postal Service.

In a University of Tennessee survey we found that 95% of all WMS users employ their WMS for the first four functions above: receiving, put away, picking, and shipping. The other functionalities are less-used. For example, fewer than 10% use their WMS for cross-docking assistance. Twenty-four percent use task interleaving. Only 35% use a mature yard management system dock management capability. Fifty percent use profiling, and 70% use their WMS to enable cycle counting and inventory accuracy.

A major challenge that some firms have is multiple WMSs. This is especially true of companies that have grown by acquisition. Standardizing WMSs is expensive and time consuming, and the ROI must be clear to undertake such a major initiative.

Many DCs cannot operate at all if their WMS goes down, so there must be a culture of 100% uptime for it. Given the scope of this advancing technology, the WMS should probably go through an upgrade every five years.

Automation

Many Lean advocates have an appropriate mantra as noted above: "eliminate before you automate." Otherwise, waste becomes hard-wired into the operation. That said, technology continues to advance, and automation systems are becoming increasingly more viable. A growing number of automated material handling systems have sensors and intelligence that optimize performance. Even forklifts are getting more sophisticated. Today's smart forklift includes diagnostics that provide alerts for required service, collision detection, fork speed optimization, speed controls for busy sections of the DC, and more.

Many processes may need to be controlled and optimized in a DC, from modern conveyor systems to automated guided vehicles and automated storage and retrieval systems. Maybe in the not-so-distant future warehouse operators will control automation using hand signals, reminiscent of Wii video games, or use Google Glass as a guide to warehouse movement. A worker at a dock might use a simple hand signal captured by digital image to trigger a conveyor. At a more basic level, a robust Wi-Fi capability will be important to support wireless devices like rugged tablets.

Because of the increase in automation, warehouse execution software is also emerging as a critical systems tool in large, modern DCs. Warehouse execution software systems provide the foundation to run automation systems, but they need to be interfaced with the WMS. The warehouse execution software must know the real time statuses of all the automated equipment in the DC such as conveyors, carousels, and picking systems, and interface them with the WMS.

In summary, thanks to recent technological innovations, the benefits of warehouse automation are becoming more accessible. Companies are increasingly automating processes such as put away, picking, sorting, and palletizing. If implemented properly, automation can reduce labor and increase productivity. Automation can also reduce cycle time and provide the capacity for strategic growth. The large up-front costs require scale, however, and automation systems can be costly to change once installed. Therefore, the ROI needs to be done realistically, and the implementation needs to be highly disciplined. That said, it is a good time to reconsider the preconceived notions about automation as technology continues to advance.

Choosing a Warehouse Management System

The annual sales volume of WMSs exceeds one billion dollars, drawing many players in the space. WMSs can be obtained from many software companies, including Manhattan, SAP, JDA/RedPrarie, High Jump, Oracle, and countless others. It is increasingly common for companies to use cloud-based systems—such as Software as a Service (SaaS)—that have been around for about a decade. In the interviews we conducted for this chapter, most of the supply chain practitioners indicated that they thought of cloud computing as "interesting, but having unclear usefulness." This is a more tepid response than what is heard from IT professionals and tech companies. These professionals tout the advantages of a cloud-based system, such as:

- Lower implementation cost
- Lack of hardware to buy
- Faster time to payback
- Lower up-front cost by avoiding a large licensing fee
- Portability of facility moves
- Automatic upgrades

Potential customers of a cloud-based system must determine the costs over the total life cycle, and these are not necessarily lower. Also, response time should be determined with a cloud-based system to avoid unproductive waiting by operators. Cloud-based systems can be extremely difficult to customize. WMSs have multiple touch points where the application must interface with other systems in the DC and outside the DC, such as the transportation management system and the enterprise resource planning system. In addition, substantial unique configuration of the software could be required to interface with sophisticated material handling equipment. Cloud based systems may make more sense for small companies who do not have the resources to deploy a WMS, or for a larger company that wants one standardized system for multiple sites.

To choose the appropriate WMS, leading companies first develop a list of requirements. Then they invite prominent WMS vendors to demonstrate how their software meets those requirements. "See it before you believe it," is a good philosophy to adopt during the sales process; software vendors have been known to exaggerate. A number of elements germane to the selection are:

- Vetted references
- Ability to integrate easily with the existing enterprise resource planning or legacy systems
- Licensing cost, and ongoing support cost
- Estimated ROI
- Scalability and flexibility to accommodate future requirements
- User friendliness
- Financial viability of the supplier
- Quality of after-hours support
- Cost–benefit analysis
- Stable software (unless there is tolerance in the firm for a beta version)
- Buy-in of the line organization

A scoring system, such as a rubric based on the above criteria, should be used for software selection. To enhance buy-in, the scoring team should be large and cross-functional. Each element in the evaluation criteria should be listed and weighted for its relative importance. For each major capability required, the scoring could be something like the sample that follows in Table 4.3:

Table 4.3 Simple Software Scoring Systems Can be Helpful

X	Not enough information
1	Does not meet needs
2	Partially meets needs
3	Meets needs but open questions remain
4	Meets Needs
5	Meets and exceeds needs

Once the weighted totals are tabulated and a good dose of judgment applied, the decision should be clearer.

Change Management

Any new technology implementation always has major change management issues with which to contend. In several surveys, we have found that supply chain professionals rank implementation difficulty as shown in Figure 4.3 below:

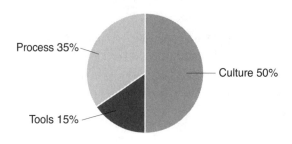

Figure 4.3 The cultural transition plan for new technology is the biggest priority

In other words, half of the implementation challenge is culture-based. Therefore, the lion's share of the challenges comes from cultural issues, not technical issues. We believe four key principles of change management should be addressed:

1. Have a plan to sustain the change. Sustaining change is harder than the initial implementation.
2. Put communication tasks in your project plan and plan the time to do it right. Getting buy-in is time consuming.
3. Manage expectations carefully.
4. Focus communication on the key individuals.

These four considerations should form the foundation for a documented change management plan.

Layout and Space Optimization

Productivity declines exponentially if DCs become too crowded and the docks, staging areas, and aisles become congested with product. In addition, overcrowding often drives a company into extremely expensive overflow warehousing. Therefore, it behooves any firm to make sure that they have given warehouse space optimization the priority it deserves.

One durable goods company told us that they had over 8 million square feet of warehousing space but were experiencing severe space shortages. They launched a number of initiatives to consolidate and averaged saving just over 10% of the space in each warehouse. That amounted to 800,000 square feet of additional warehouse space, which they avoided having to pay $32 million (or $40.00 per square foot). What were the initiatives that this company used to optimize so much warehouse space? Below are some ideas that worked for them and that have worked for others.

Eliminate Honeycombing

Honeycombing is a symptom of too many slow-moving SKUs, which consume their space in a sporadic and partial manner, creating an uneven, "honeycomb" look to the stacks in the warehouse. Leading companies group or consolidate slow-moving or obsolete SKUs in remote areas of the warehouse. They find that combining slow movers in the same area takes a modest amount of additional labor but saves major amounts of space.

Racking and Mezzanines

If products are not stackable, then racks and mezzanines can be used to effectively fill the cube of a warehouse. Racks need to be safe. Single deep racks not anchored to a wall or columns are inherently unstable. Lag bolts to the floor are helpful but not adequate. Back-to-back rack installation is much more stable and thus safer. The prime risk to racking is accidental collision with forklifts.

Warehouse racking should also contain sufficient capacity to handle every SKU that cannot be stacked. DCs should evaluate the feasibility of mezzanine space over every area that does not fill the warehouse cube.

Stacking Height and Product Stackability

Stacking heights should be challenged if the warehouse cube is not full. One DC had one-ton bags of a chemical product stacked in racks three high. In addition, they had a major amount of overflow product stored in trailers in their yard. This generated a huge cost for trailer rental. Later, they realized that three-high stacking was simply a traditional paradigm dating back to the warehouse's inception rather than safety protocol. They moved to four-high stacking, which increased the size of the warehouse by 33%, and eliminated millions in overflow storage cost.

Sometimes product packaging inhibits the ability to stack product in a warehouse. Product design engineers, in their zeal to reduce cost, can make it impossible to stack product in a warehouse just to save a few cents on cardboard. This wastes huge amounts of DC cube.

Packaging needs to be even more robust in some environments like warm climates with high humidity.

There are a few other things to consider with stacking height and product stackability. Stacking may be inhibited by local statutes, like fire codes. As noted above, strategic use of mezzanines can better utilize the warehouse cube, especially when placed over nondistribution functions. Warehouse space above 30 feet is likely inefficient to retrieve with lift trucks and may best be served with an automated retrieval system.

Eliminate Nondistribution Functions

Nondistribution functions are often carried out in a DC, such as:

- Customization of units with unique branding, labeling, packaging, final assembly, and kitting
- Postponement activities
- Handling of returns, refurbishment of damage, repackaging, and storage of components to support these functions
- Using maintenance areas, and space for lift truck parking and battery charging, as well as maintenance supplies
- Offices

Step one is to document the space and warehouse cube allocated to activities such as these. Then ask if there is a better way that wastes less of the DC cube. For example, these activities might be placed in or under mezzanine areas to better consume the DC cube.

Warehouse Layout: Docks and Aisles

How wide should aisles be? If fourteen-foot aisles are traditionally used, perhaps twelve-foot aisles should be considered. Aisles take up a major amount of DC cube. Aisles should be as wide as necessary, *but no wider*.

Many software tools are available, from sophisticated computer aided design systems (i.e., SIMCAD) to tools anyone can download from the Internet. Some can produce a three-dimensional, digital layout of the warehouse, and then model and optimize space based on a set of constraints. Various layouts can be simulated and animated. Three-dimensional simulations can be built to allow managers to identify problems before they are hardwired in place.

Receiving and Shipping

Receiving and shipping functions, although critical to the efficiency of the overall warehouse operation, often have either too much or too little space allocated to them. Some up-front analysis is in order. To properly design the receiving and shipping areas, one must know the following:

- Type of material to be shipped and received
- Frequency of activity now and in the future
- Nature and numbers of vehicles
- Types and numbers of docks
- Staging area required
- The Pareto distribution (i.e., 80% of the activity will be generated by 20% of the items)

With such data, leading companies can use a tool to simulate dock activity to make sure that any bottlenecks are identified and the proper amount of space allocated.

Network Optimization

Several supply chain professionals have told us, "If you have not completed a network optimization study recently, you are probably leaving a major savings opportunity on the table." We agree. A network optimization answers questions such as:

- How many DCs should you have?
- Where should they be located?
- Which customers, stores, or locations should they serve?

Many retailers routinely tweak the stores served by each DC with multimillion-dollar savings annually. Both retailers and manufactures can evaluate and change the structure of their DC network to produce major savings. Some firms do a network refresh on a periodic schedule, like every two to five years.

Some very powerful software tools are available to assist in this endeavor. Most are optimizers, which can minimize cost while satisfying a series of constraints, such as those from customer service. Some also use a simulation capability to accommodate more extensive and realistic data sets, as well as probabilities. Optimization studies can get very large for a complex network. Some companies attempt to consider a wide range of variables, such as the location of suppliers, customers, and factories. One company's model included a mind-boggling 12,089 variables, 58,552 constraints, and 2.9 trillion input parameters. Even with modern computing power, the data set on optimization studies must often be simplified. Fortunately Moore's Law—the idea that computing power doubles every two years at no additional cost—is gradually resolving that problem.

Llamasoft is a good example of a leading provider of network optimization software. Most companies use outside expert assistance, such as consultants or a 3PL, when they conduct studies like this. A few leading companies have in-house teams often staffed with people who have industrial engineering backgrounds. One company has a supply chain engineering department, another has a global analytics group, and yet another has an in-house strategy group to perform such studies. Many companies do a network refresh study on a routine basis, approximately every two to five years. In between, analytical people can also do quick and dirty studies effectively if the scope is limited.

Many organizations are reevaluating their networks in light of the Internet boom. They study whether website orders should be served out of stores, regular DCs, or specialized DCs.

A network optimization study is as much art as it is science. A great deal of management judgment must be applied to successfully complete an implementable plan. And of course, once a general location for a DC is selected, many other factors come into play to select a specific site, such as:

- Proximity to manufacturing plants
- Distance to market/customers
- Real estate/leasing conditions
- Local cost of living
- Available workforce, skills and flexibility
- Local transportation, both inbound and outbound

Network optimization studies are traditionally difficult and quite resource intensive. They are challenging for many reasons:

- Organizational boundaries are crossed
- Data requirements are massive
- Often the data must be corrected (most companies have extensive errors in their data)

- Projects are large and require dedicated resources
- Uncertainty exists in many variables
- Implementation requires extensive change management

A good rule of thumb is that assembling and cleansing the data will take 70% of the project time. As one experienced executive noted, "It will take at least twice as long as the most pessimistic person thinks." Management of scope for such projects becomes critical.

There are often unintended positive benefits to a network optimization study. For example, to feed the model, one analyst needed to know how many units a certain small boat manufacturer loaded in each truckload. When he learned the answer was two boats in each truck, he naively asked, "Why so few?" That simple question motivated people in the firm to challenge the paradigm and eventually increase that number to six units per truck. At another company, a large number of errors had been discovered in the data and many numbers had to be corrected for the model. These data were used in other applications for important company decisions. By cleaning the data, other applications also performed much more accurately.

Safety and Security

When asked for their highest priority, warehouse professionals more often than not say safety is number one. Safe DC operations require a combination of mindset/culture, metrics, and processes.

Safety Mindset/Culture

Everyone in the facility every day needs to think safety. That starts at the top when senior executives state openly and often, "safety is our number one priority." Norfolk Southern Railroad has a rule: *Every meeting in this company will begin with a safety briefing.* Another firm calls this a "safety share" at the beginning of every meeting. Although most would consider that overly tedious, it sends a message to every employee, every day. Other companies feel they can make the point with monthly focused safety meetings. Some firms put safety in their corporate mission statement. Some state it clearly in their corporate values. Some have an official safety strategy that they update every year.

A large global manufacturer we visited requires employees and visitors to watch a safety video, and puts employees through regular safety training. Once completed, the visitor receives a card that must be shown each time entering the facility. A safety briefing is appropriate for all visitors to DC facilities, including basic information such as how to exit the building in an emergency. One firm briefed visitors on the way to avoid any toxic fumes, telling them, "When you exit, look at which way the flags are blowing and proceed up wind."

Safe operations relentlessly require people to wear the proper clothing, including safety glasses, hearing protection, orange vests, and gloves, sleeves, and steel-toed shoes when appropriate. Visitors on a simple tour are no exception to these dress code rules.

As mentioned earlier, Lean operations often add a sixth "S" for safety to their 5S process, further enhancing the safety mindset. One corporation has a safety wall in each DC. These are large, twenty-five-foot long displays showing safety stats and celebrating milestones. Another company has large safety celebrations when they reach a milestone with no lost time injuries. They said they spend $25,000 per DC on a cookout, where they give away merchandise. One firm encourages associates to put up pictures of their loved ones, employing a "Do it for them" campaign. All of these actions contribute to a culture of safety.

Actions speak louder than words. It is never acceptable to sacrifice safety when some other objective becomes paramount. Labor productivity, output and/or quality are all critical to survival, but employees can never see safety take a backseat to any of them.

Temporary employees represent more of a challenge. One company told us that temporary employees experienced reported safety problems 40% higher than regular employees. When asked why, the answer seems to be a combination of too little vetting (drug screening, previous work history) and too little training.

Safety needs to be everyone's responsibility. One company selects a safety advocate for each facility. The job can be done on a rotation basis, and the individual should be accountable to oversee training, metrics, and safe processes. Another firm has a safety committee in each DC led by an hourly associate. They believe safety is all about total employee involvement and ownership. This firm required all 10,000 of its employees to take five modules of safety training. Leading companies understand that safety cannot be punitive or else the associates will not participate. The hourly associates must own safety and aggressively report unsafe acts.

When considering safety, the subject of hazardous materials cannot be ignored. A material safety data sheet provides workers and emergency personnel with procedures to safely work with that substance. Material safety data sheets should be kept up to date with information such as physical data (melting point, boiling point, flash point), toxicity, health effects, first aid, reactivity, storage, disposal, protective equipment, and spill-handling procedures.

Safety Metrics

Safe operations have visible metrics along with serious accountability for them. Common safety metrics include:

- OSHA recordables and lost time events
- Corrective action completed to address safety problems
- Safety training completion
- Safety survey results
- Time since last incident

Leading corporations track these metrics over time and make them highly visible throughout the operation.

Tracking the most common and the most severe injuries to determine their causes is important. For example, the most common injuries may be musculoskeletal injuries caused by lifting, pushing/pulling, slips, trips, and falls. The most serious injuries often occur when someone is caught between or hit by equipment. A fatality is horrific and often results from what is initially perceived as a freak event. One company had a fatality when someone simply jumped off the end of a truck dock. At the root of any such horrible occurrence is the opportunity to make fundamental safety changes.

Safety Processes

Safe operations have safe processes. Safety audits should be performed on a periodic basis. Organizations should be relentless about meeting and exceeding all OSHA standards. A few things to look out for include:

- Unsafe use of forklifts (people need to be separated from lift trucks by guard rails whenever possible)
- Electrical dangers
- Improper ventilation or handling of hazardous materials

- Improper stacking
- Failure to use personal protection equipment
- Failure to have strict lockout/tag out procedures for equipment being serviced
- Inadequate emergency systems for fire or injury like chemical burns
- Improper storage and handling of hazardous materials
- Improper procedures for repetitive motion or lifting
- Unprotected pinch points (around conveyors, for example)
- Bad general housekeeping (cluttered aisles and docks)
- Failure to block or protect dock doors
- No cages over ceiling lights to prevent glass from dropping
- Failure to adequately prevent explosion risks (like smoking around charging stations)
- Poor lighting
- Failure to protect pickers in stacks with enclosed cages or a failure to continually wear a harness on high-rise vehicles

Security

Security to avoid theft and pilfering is more important for some products than others. Truckload value does not have to be astronomical to merit security measures. Dishonest employees can defeat the most elaborate security system for high value items or high theft items like electronics, liquor, or drugs. That is why front-end screening must be very aggressive in such facilities, even including polygraph or voice-stress analysis. High theft items must be placed in their own secure areas.

Only certain employees should be able to enter secure areas and/or break seals. The temptation grows once cartons are opened. Modern, hard-to-defeat electronic alarm systems are a must. High fences and guard dogs around a warehouse perimeter help. There should be a rigorous sign-in/sign-out procedure and a badging program. High-resolution cameras, metal detectors (for everyone, even visitors, and the boss), and turnstiles can be essential security devices where appropriate. There should only be one way in and out of a DC, barring an emergency situation, and that entrance/exit should have detectors. Trailer yards can also be a weak point. Tractors and trailers have been stolen from secure fenced-in areas.

Companies should hire a good security expert to oversee security systems. One company has loss prevention personnel on site around the clock to man cameras, administer entries and exists from the facility, and inspect outbound trailers. Sources may be people with backgrounds in government agencies like the FBI, Secret Service, or military police, or people who have worked for security firms like ADT.

On a global basis, the Customs-Trade Partnership Against Terrorism (C-TPAT) is a voluntary supply chain security program led by US Customs and Border Protection. It focuses on improving the security of private companies' supply chains with respect to terrorism. The program was launched in November 2001 with seven initial participants, all large US companies. Today, the program has over 10,000 members.[2] The participants in the program account for well over half of all merchandise imported into the United States. Companies who achieve C-TPAT certification must have a documented process for determining and alleviating risks throughout their international supply chain. These companies are considered low risk, so they receive expedited cargo processing and fewer customs examinations.

People

Having the right people in the right jobs is key. Leading companies excel in five areas when they manage talent, and their DC operations' talent management should be no exception. The five steps in a good talent management process are: assess needs, identify job candidates, hire, train and develop, and retain. These principles are further detailed in Chapter 6.

As was mentioned in the section on Lean warehousing, it is essential that all associates in a warehouse operation be engaged and involved in continuous improvement. Some companies have success with self-directed workforces. The supervisors become problem-solvers, and the associates take on routine supervisory duties—including hiring, work scheduling, and job rotation. Although more challenging, this has also been successfully implemented in unionized environments.

Leading companies invest in training for new employees as well as ongoing training. Leaders also find a way to retain good employees

---

[2] "C-TPAT: Customs-Trade Partnership Against Terrorism," *U.S. Customs and Border Protection, Border Security*, http://www.cbp.gov/border-security/ports-entry/cargo-security/c-tpat-customs-trade-partnership-against-terrorism (retrieved May 2015)

and minimize turnover; they pay a fair wage for the area, show appreciation, and foster an atmosphere of ownership by involving employees in DC improvements.

Third-Party Logistics Operators

An increasing number of companies use third-party logistics to manage their DC operations; this includes hiring and managing the hourly associates. There is a continuing trend toward more firms outsourcing their DC operations to 3PLs. Third-party logistics firms can do many additional things, such as:

- Reduce future cost by leveraging expertise and technology
- Improve customer satisfaction
- Provide a contingency plan for union issues
- Reduce risk
- Reduce current cost
- Provide global expertise, including documentation, customs, and duty optimization

Great people, management and hourly associate personnel are essential to accomplishing those goals. The survival of a 3PL depends on the caliber of its employees. Many 3PLs partner with an employment company. These partner agencies thoroughly vet personnel and provide some training.

Hourly Associates

Often, new hourly associates are temporary workers for ninety days before becoming full-time employees. Sometimes the firm or 3PL pays less than competitive wages during this initial period. This practice is counterproductive in that it often leads to more turnover, which can be over 100% for temporary workers, and fewer ideas for improvement brought forward by the associates. Of course, higher wages do not automatically lead to better results. They must be coupled with the right culture and processes.

Firms that develop and retain great management talent have a clear advantage over those who do not have a formal plan. One professional observed that many professionals ask, "What if we train and develop our management team and hourly associates and they leave?"

His retort, which we believe is the correct question, "What if we do not develop them and they stay?'"

Many DCs need an extremely flexible workforce that is comfortable with highly variable working hours. The work mirrors the shipping pattern, and often shipping/demand patterns are highly variable.

Kaizen

Hourly associates always provide the best source for kaizen, or continuous improvement, as discussed in the section on Lean. One company instituted a structured suggestion program and received over 500 employee suggestions in nine months. The suggestions lead to savings of hundreds of thousands of dollars and countless safety improvements. Supervisors need to focus on being problem-solvers rather than bosses. Sadly, not all supervisors can get with this program. Another company noted that, if you do not involve people, they will focus their substantial brainpower on other things, and they will work very hard to not work hard.

Management Personnel

Each professional in the organization should have a personalized development plan that is tracked. Some companies invest heavily in their in-house development. For example, Kenco has a leadership development program with quarterly one-week training sessions over an entire year. The great majority of the program's graduates have stayed with the company and have been promoted. Kenco also annually sends two employees with high potential to the University to Tennessee Global Supply Chain executive MBA program. Modern executive MBAs like these can cost upwards of $100,000, but they have been proven to generate ample ROI for the sponsoring company.[3] Progressive corporations view people as an asset, not an expense. Not only are people an asset, but also they can and should be an appreciating asset. Even in tight times, a training budget needs to be liberal and protected.

---

[3]*Graduate and Executive Education Department at the University of Tennessee Haslam College of Business* (June 2014).

Sustainability

The green revolution is upon us. More often than not, the supply chain is at the forefront of these organizational changes. The extended supply chain generates much of the carbon emissions for most firms, so it makes sense for the supply chain team to lead the charge on corporate sustainability. Warehouse operations are key to an overall supply chain sustainability effort. In one large company, the senior vice president of supply chain reports directly to the corporate executive vice president of sustainability.

DCs can do many things to be green, and fortunately many of them have a financial payback. Most supply chain professionals pursue green initiatives to cut costs, although public relations and brand image can also play a role. Some common green initiatives are:

- Cardboard recycling, which can turn into a lucrative business for some companies
- Plastics recycling and minimization of shrink-wrap usage
- Moving from propane to electrical lift trucks (The next step may be fuel cells that run on hydrogen and emit water vapor. Today the ROI on the required cost and infrastructure is problematic, but that should change over time. Many companies are running pilot programs with fuel cells.)
- Working with vendors to reduce packaging materials (Some DCs must have large trash-cutting operations to remove excess packaging. A large industry consortium has been formed with the goal of eliminating inner packaging material, working with the outstanding packaging engineering program at Michigan State University. Twenty-five companies recently participated in a conference on that topic.)
- Returnable and collapsible containers
- Reuse of wood pallets, even making mulch with worn-out pallets
- Caulking and weather stripping around windows and doors, seals around dock doors, and appropriate insulation
- Automatic adjustments to the HVAC system when the building is not in operation
- Natural lighting such as skylights or solar panels. One DC in California said that they have solar panels covering the entire 700,000-square-foot roof. Several firms said that solar panels have a very low ROI, but "seem like the right thing to do"

- Motion detectors for lighting in low traffic areas and photo sensors for outdoor lighting
- Less lighting where it is not needed
- Efficient lighting systems. Lighting is the largest energy cost in most DCs (One warehouse professional we interviewed told us that LED fixtures with intelligent integrated controls, sensors, and reporting can save an estimated 50% of lighting costs. There can also be both local and federal tax incentives for projects like this. The payback on the investment is generally about eighteen to twenty-four months. The ROI depends on a range of factors such as number of shifts, ceiling height, the local cost of electricity, incentives, and activity in the various zones of the warehouse. Intelligent motion sensors also save lighting cost. After no motion for a specified time the light can dim and then later shut off completely.)

Often lighting is the easiest energy issue to address, but it is important to track all energy—not just electricity. Electrical usage, natural gas, propane, water, and sewage should all be tracked and monitored, with usage goals set for each.

Summary

The incredible scope of DC management requires the best and brightest of management personnel. This job demands innovation, technical acumen, and human resource skills. DC managers must manage upward to respond to very tough management goals and customer expectations. They must also manage downward to engage all employees in a journey of continuous improvement. The relentless challenge for companies is to find and retain talented people who can manage the incredibly difficult, ever-changing DC operations of the future.

We have included a DC management evaluation tool in Table 4.4. This tool can be used to evaluate where you stand versus the best practices described in this chapter.

About the Global Supply Chain Institute

http://globalsupplychaininstitute.utk.edu/

The Global Supply Chain Institute at the University of Tennessee has one of the premier supply chain programs in the country. UT

Table 4.4 Effective Assessment Tool for Your Warehouse Best Practice Capability

Distribution Center Evaluation Tool

Note: This tool applies only to larger DC operations

Please rate each item on a 1–10 scale, where 10 is world-class

Receiving		Lean Warehousing	
1. Do you receive ASNs on most of your inbound receipts?		9. How advanced are Lean philosophies and processes in your DCs?	
2. Do your suppliers deliver high quality shipments on time, with accurate counts?		10. Do you heavily involve the hourly associates in your improvements efforts?	
3. Do you have a robust process to manage returns?		11. Do you have a robust Six Sigma process in place?	
4. Do you quickly put product away and in a place best for picking?		**Cross Docking**	
		12. Do you have a mature cross-docking process, appropriate for your operation?	
Picking/Order Fulfillment/ Shipping		**Metrics and Planning**	
5. Do you have an excellent profiling/ slotting process to put SKUs in the best location for picking?		13. Do you have a documented, multi-year strategic plan for your DC operations?	
6. Do you have an appropriate level of automation and systems to accurately and efficiently fulfill orders?		14. Do your metrics have a logical framework, along with appropriate goals?	
7. Do you have an excellent process in place to fill Internet orders?		15. Do you have a balanced set of metrics covering safety, people, customer service, cost, and asset management?	
8. Do you maximize the cube when you load trailers?		**Warehouse Management Systems**	
		16. Do you have a modern WMS in place?	
		17. Do you effectively use an appropriate level of the WMS's functionality?	

Table 4.4 (*Continued*)

Layout and Space Optimization	
18. Do you effectively utilize the full cube in your DC locations?	
Network Optimization	
19. Do you do a network optimization study at least every five years to refresh your supply chain network?	
Safety and Security	
20. Do you have an intense safety mindset and culture?	
21. Do you have visible, impactful safety metrics?	
22. Do you have safe processes?	
23. Is your DC secure from pilferage?	

People	
24. Do you have a complete talent management process in place for your DC organization including the five elements of: assess, identify, hire, develop, and retain?	
Sustainability	
25. Do you have an environmental strategy in place for your DC operations?	
Total Points	

How did you do?

0–100 points: Your DC operation is deficient. You should benchmark best in class operations and develop a multi-year plan to upgrade your operation.

100–150 points: You have an average DC operation. You have a solid position to build on. You now need a multi-year plan to improve your DC operations.

150–200 points: You have a good to excellent DC operation. Build on your many strengths. But also honestly assess and address your weaknesses.

200+ points: You have an outstanding, approaching world-class DC operation. Make sure you keep it that way. Everyone is raising the bar everyday. Keep challenging yourself to remain at the top.

offers undergraduate, MBA, and doctoral student education within the field, and its thirty supply chain faculty members are ranked No. 1 for academic, supply chain research productivity.

Transition: The ABCs of DCs— Distribution-Center Management: A Best Practices Overview to Bending the Chain—The Surprising Challenge of Integrating Purchasing and Logistics

Distribution-center management is no easy task for supply chain professionals. The eleven themes covered in Chapter 4 are topics that must be addressed if a DC is to function at its maximum potential. Fortunately, there is a set of best practices for each topic that managers can use to evaluate the effectiveness of their own DCs.

A critical component of distribution-center management is inventory. Inventory "feeds the beast," per se. How a leader manages the inventory in their DCs determines the success of the supply chain, from getting product delivered to the final consumer to saving money on the bottom line. But effective inventory management cannot be accomplished if functional supply chain areas are not aligned. When procurement and logistics are not operating together—with clear direction and purpose—the entire supply chain is losing value.

This concept of purchasing and logistics integration is covered in Chapter 5: *Bending the Chain—The Surprising Challenge of Integrating Purchasing and Logistics*. Bending the chain, or uniting these two functions, is a measure that gives best-in-class companies their edge. Even though the unification of procurement and logistics is correlated with financial success, the firms with which we have consulted suggest they have made little progress toward this type of alignment.

This chapter is meant to educate supply chain professionals on how best-in-class companies have aligned their purchasing and logistics functions. It explains just how challenging this integration can be, and it explains the research that links purchasing and logistics integration (PLi) to improved functional and financial performance.

We have found that elite companies use four best practices to successfully bend the chain. This conclusion comes as a result of extensive research consisting of interviews with leading supply chain firms.

After the best practices are established, we present a 7-step strategy leaders can employ to make a difference in their own supply chains.

This chapter is also complete with a PLi assessment. The worksheet can be used to determine just how good your firm is at integrating purchasing and logistics. Key members of your team should take the assessment so that you might see where their answers match and where they differ. You should expect to learn quite a bit about your supply chain, including opportunities to improve your processes.

Even though purchasing and logistics are both supply chain functions, oftentimes they are not on the same page. This must change if supply chains are expected to prosper when challenged with the game-changing trends addressed in the first chapter. Bending the chain is one of the best ways to make this happen.

5

Bending the Chain: The Surprising Challenge of Integrating Purchasing and Logistics

A Report by the Supply Chain Management Faculty at the University of Tennessee

Ted Stank, Ph.D., J. Paul Dittmann, Ph.D.,
Chad Autry, Ph.D., Kenneth Petersen, Ph.D.,
Mike Burnette, Daniel Pellathy, Ph.D., candidate

Executive Summary

Over the last several decades, supply chain (SC) professionals have focused on performance issues that have emerged from a lack of commercial/business alignment with supply chain operations. Significant improvements have been made, and systemic processes such as integrated business planning (IBP) and sales & operations planning (S&OP) have been developed to drive a fully integrated business. As business integration has continued to improve, the biggest supply chain opportunities have shifted.

Every year, the University of Tennessee's Global Supply Chain Institute networks with hundreds of companies, requesting information on emerging supply chain issues. Our recent research shows that one of the greatest business integration opportunities is found within the traditional supply chain functions themselves. Specifically, we believe a major strategic integration opportunity exists between purchasing and logistics, and failing to capitalize on this opportunity is causing many firms to miss important opportunities to create value.

Based on our research, we believe it is probable that your firm is organized, measured, and incentivized in ways that essentially prevent you from deriving the full benefits of collaboration. In fact, it is highly likely that your company encourages behaviors that destroy value, both in the short term (by sub-optimizing total system costs) and in the long term (by generating superficial gains from functional cost reductions while failing to leverage asset investments).

We have also uncovered strong evidence that organizations that align procurement and logistics decisions vertically with business unit strategy and horizontally between functions enjoy heightened levels of both functional and financial performance. In essence, these high-performing companies are able to **bend the chain** of plan, source, make, and deliver to enable alignment between purchasing and logistics. The result is that they serve customers better with lower operating expenses, cost of goods sold, and inventory.

Our research also sheds light on the structures, processes, and tactics top firms employ to enable this type of functional integration. Data from over 180 supply chain leaders (firms ranging in size from over $20 billion to under $100 million) were collected and have allowed us to draw the following high-level conclusions:

- Procurement and logistics frequently are found in a broader supply chain or operations organization but really exist as two separate and disconnected functions.

- Both procurement and logistics are well-aligned independently to their business unit's strategy and activities but not nearly as well-aligned to each other.

- Despite formal organizational links between purchasing and logistics, interaction between these functions is typically informal and unstructured.

Similarly, in our own experience we have found that when functional elements of the supply chain align with each other, improvements in firm financials and earnings per share invariably follow. Without integrated decision-making, financial performance is at best sub-optimized and at worst destructive to value. Firms must refocus organizational design, metrics, talent, and incentives to align activities across the value chain.

Finally, we conducted an analysis to determine whether the data provided any indication as to whether procurement and logistics integration (referred to as PLi in this paper) was perceived as being an important lever of overall business success. The data clearly show that *integrated purchasing/logistics organizations deliver better business results* (i.e., cost productivity, working capital productivity, and product availability).

Additionally, the many interviews we conducted with leading supply chain firms clearly suggest that companies with "best-in-class" supply chains consistently deliver the strongest business results. These best-in-class organizations tend to employ a set of four best practices:

1. Fully integrated end-to-end supply chain organization integrated with common metrics.

2. A talented supply chain organization that rewards people for in-depth mastery and end-to-end supply chain leadership.

3. A purchasing and logistics network with an operating decision framework based on best overall total value of ownership (TVO; TCO plus level of customer value creation).

4. Effective information systems and work processes that enable superior business results by providing multifunctional supply chain teams with the proper tools and information.

Finally, through our research and best-in-class interviews, we have been able to define a short list of actionable steps supply chain leaders can take today to make a difference.

1. *Get it on business leader scorecards.* Change the business reward system and culture from "sub-optimal functional goals" to "total value creation for the enterprise."

2. *Champion TVO.* It is not enough to talk use of total value of ownership with your direct reports. Personally lead the change in the supply chain.

3. *Make R&D your best friend.* Create a seamless technical community that is aligned on total business value creation between R&D and supply chain. New product supply chain design should be a seamless technical community deliverable.

4. *Set clear expectations on the use of multi-discipline teams in analysis and decision-making.*

5. *Champion an end-to-end and integrated supply chain organization.* In the short term, align on a common direction if

the purchasing and logistics teams have different leadership. Ensure that both organizations have a common supplier direction, scorecards, and rewards.

6. *Build supply chain talent that includes end-to-end supply chain mastery*.

7. *Partner with finance*. Work with finance leadership to align on how your multi-discipline teams quantify value for quality, customer service, environmental, sustainability, delivery, cost, and inventory.

The Surprising Challenge: Purchasing and Logistics Integration

Quality management icon W. Edwards Deming asserted over 30 years ago in the first of his famous 14 points that a business enterprise needs constancy of purpose to succeed. Without this consistency of purpose, the business is not an organization but rather a collection of functions acting in disjointed and contradictory ways, impeding or even destroying value. Obvious improvements cannot be implemented, and ultimately business activities fail to create a chain that produces value for the company and its customers. Deming's solution to this fundamental problem was to focus on overt collaboration between functions. However, as our research shows, most companies still fail to follow through on his prescription.

Instead of adopting this advice, all too often organizations have focused on developing technical centers of functional expertise to drive scale and meet short-term financial and market expectations.

In the past five years, we have conducted over 700 interviews with managers across all industries as part of the University of Tennessee's College of Business supply chain audit program. At the end of every interview, we always ask a "wish list" question: if you could change the world, what would you do to improve things in your company? By far the most common answer to that question is the desire for all of the functions in the company to work together and become perfectly aligned toward a common purpose. People we interview pine for an environment where the functional silo walls have come down. They intuitively know that these disconnects are the real reason things are not improving faster.

In this chapter we discuss the results of a large-scale research initiative, along with real-life industry examples, which point to the fact that collaboration across functions and between enterprises is

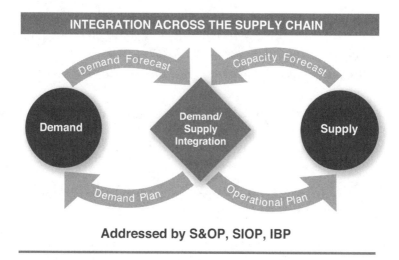

Figure 5.1 Integrated supply and demand drives revenue and cost results

woefully missing from the value chain practice despite at least three decades of focus in the popular and academic press. More importantly, we show that when processes are integrated and silo walls are eliminated, the results can be very significant. Figure 5.1 is a basic interpretation of demand/supply integration across the supply chain.

As research by consulting firm Oliver Wight has already shown,[1] when companies integrate, the following changes result:

- Revenue goes up 10% to 16%
- Fill rates go up 10% to 48%
- Logistics costs go down 10% from 32%
- Inventory goes down 15% from 46%

Similarly, in our own experience we have found that when functional elements of the supply chain align with one another, improvements in firm financials and earnings per share invariably follow. Without integrated decision making, financial performance is at best sub-optimized, and at worst destructive to value. Firms must refocus organizational design, metrics, talent, and incentives to align activities across the value chain.

---

[1]Oliver Wight International, Inc., *The Oliver Wight ABCD Checklist for Operational Excellence*, 5th ed. (New York, NY: John Wiley & Sons, 2000).

Supply and Demand Disconnects

For years, supply chain leaders have debated and discussed the disconnect between the supply and demand sides of business organizations. This lack of integration between sales and operations has spawned entire industries around ideas like S&OP, sales, inventory, and operations planning (SIOP) and integrated business planning (IBP).

But the disconnects go far beyond this macro level. For example, there is often a lack of integration within the demand side of the firm. Sales and marketing in manufacturing companies are not always aligned, nor are sales, marketing, and merchandising in retailers. Figure 5.2 visualizes this disconnect.

Supply Side Disconnects

There are similar disconnects on the supply side among logistics, operations, and procurement. Figure 5.3 builds on Figure 5.2 by introducing these disconnects on the supply side. Our research shows that one of the greatest opportunities for "lack of integration or dis-integration" lay within areas traditionally thought of as supply chain functions. Specifically, we believe a major strategic integration opportunity exists between purchasing and logistics, and failing to capitalize on this opportunity is causing many firms to miss important opportunities to create value.

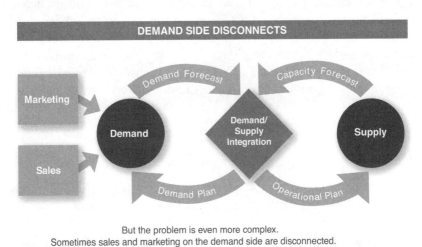

But the problem is even more complex.
Sometimes sales and marketing on the demand side are disconnected.

Figure 5.2 Sales and marketing are frequently mis-aligned

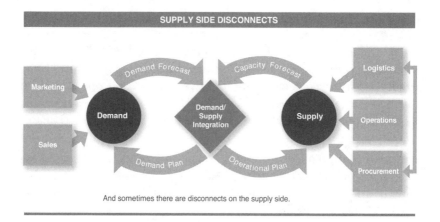

Figure 5.3 Disconnects on the supply side of the supply chain can be just as disruptive as those on the demand side

The Surprising Gap Between Purchasing and Logistics

Ideally, the supply chain functions of plan, source, make, and deliver (Figure 5.4) are aligned and focused on serving the customer while simultaneously delivering world-class cost and working capital levels. The two functional areas of purchasing and logistics each have a major impact on these goals. Together, purchasing and logistics can represent up to 70% of total organizational costs and influence 80% of working capital through inventory and payables. Yet decisions made in these two areas are rarely made in concert with each other. In fact, purchasing often focuses decision-making on optimizing metrics associated with purchase price and cost of goods sold, while logistics is focused on optimizing metrics associated with delivery and storage efficiency and effectiveness. Neither area tracks performance to higher-level financial value creation.

Example:

A logistics executive for a large global consumer durable goods company hosted a "supply chain management advisory board." During

Figure 5.4 The traditional model for supply chain excellence

dinner at a local restaurant, the executive leading this group noticed another group from their firm with a group of visitors in another private room in the restaurant. It turned out this other group consisted of the company's purchasing executives hosting their own "supply chain management advisory board." Neither group, to their collective surprise and chagrin, had any knowledge that the other group was meeting, nor did one know what the other was talking about.

What is the takeaway from this story? As both sides thought about it, they realized that it was symptomatic of a purchasing group making decisions about purchasing locations globally with no insight into costs of movement. At the same time, the logistics group was focused on how to reduce costs of global warehousing, inventory, and transportation with no insights into future locations of supply and manufacturing. The ideal plan-source-make-deliver model morphed into a new disconnected reality.

Based on our research, it is probable that your firm is organized, measured, and incentivized in ways that prevent you from deriving the full benefits of collaboration. In fact, it is highly likely that your company encourages behaviors that destroy value, both in the short term (by sub-optimizing total system costs) and in the long term (by generating superficial gains from functional cost reductions while failing to leverage asset investments).

We have also uncovered strong evidence to suggest that organizations that align purchasing and logistics decision-making *vertically* with business unit strategy and *horizontally* between the functions enjoy heightened levels of both functional and financial performance. In essence, these high-performing companies are able to **bend the chain** of plan, source, make, and deliver to enable alignment between purchasing and logistics. Figure 5.5 does a good job of visualizing this alignment. The result is that they serve customers better with lower operating expenses, cost of goods sold, and inventory. Our research also sheds light on the structures, processes, and tactics top firms are employing to make this happen.

The Research: Linking Purchasing and Logistics Integration (PLi) to Improved Functional and Financial Performance

A survey was sent to purchasing and logistics managers from the University of Tennessee Global Supply Chain Institute and Forums

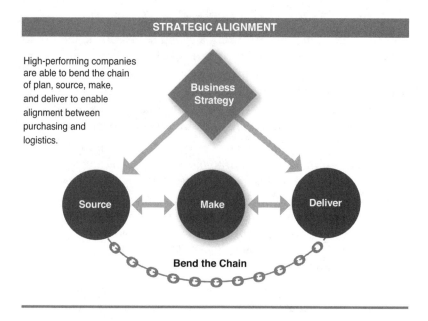

Figure 5.5 Bending the chain to drive logistics and procurement alignment

mailing list, resulting in over 180 responses from managers, ranging from CEOs and presidents to analysts. The respondent firms ranged in size from over $20 billion to under $100 million in revenue. The industries included the following:

- Aerospace/defense
- Apparel/textile
- Automotive
- Building materials
- Chemical, oil, and gas
- Commercial printing
- Components and systems
- Conglomerate
- Construction
- Consumer electronics
- Engineering
- Environmental services

- Facilities management services
- Financial institutions—banking
- Financial institutions—insurance
- Food, beverage, and nutrition
- Food service
- Government—national
- Government—local
- Health care delivery services
- Heavy machinery
- High-tech network infrastructure
- Hotel/hospitality
- Household, personal care, and cosmetics
- Industrial equipment
- Media/entertainment
- Medical equipment
- Metals/glass processing
- Mining
- Office equipment
- Packaging
- Pharmaceuticals
- Plastics processing
- Professional/information services
- Pulp and paper
- Retail
- Telecommunications services
- Transportation services
- Utilities
- White goods

Respondents were first asked to identify whether they worked primarily in purchasing or in logistics. Purchasing was defined as including the following:

- Sourcing direct materials
- Procurement of maintenance, repair, and operating supplies

- Contracting services with outside suppliers
- Procurement of capital equipment/facilities
- Procurement of finished goods (completed items for resale)
- Supplier evaluation and selection
- Management of continuous supplier relations
- Supplier performance measurement
- Establishment of goods/services specifications
- Contract negotiations over materials supplies/services
- Global sourcing/sourcing strategy

Logistics was defined as including the following activities:

- Inbound/outbound transportation
- Owned fleet management
- Warehouse operations management
- Materials handling
- Packaging
- Order fulfillment
- Logistics information systems management
- Inventory management
- Management of third-party logistics services providers
- Customer service
- Reverse logistics flows
- Supply/demand planning

Next, the respondents were asked a series of questions related to their perspective on the nature and level of integration between their department and overall business strategy as well as between the purchasing and logistics functions. For example, if respondents indicated they were purchasing managers, they were asked about the purchasing group's alignment with business strategy and the group's relationship with the logistics group.

Major findings from the survey include the following:

1. Purchasing and logistics frequently are found in a broader supply chain or operations organization but really exist as two separate and disconnected functions.

2. Both purchasing and logistics are well-aligned independently with their business unit's strategy and activities but not nearly as well aligned with each other.

3. Despite formal organizational links between purchasing and logistics, interaction between the functions is typically informal and unstructured.

4. Maintaining open lines of communication is the most widely supported method of interaction between the functions.

More detail on these findings is provided in the tabular breakdowns that follow.

Major Finding 1: Purchasing and Logistics Frequently Are Found in a Broader Supply Chain or Operations Organization but Really Exist As Two Separate and Disconnected Functions

While nearly 58% of respondents reported that purchasing and logistics were part of a common supply chain organization, over 45% felt that they exist as separate functions. Fourteen percent still viewed purchasing and logistics as separate functions that are not part of the same supply chain organization, and 28% reported some other organizational structure. Table 5.1 organizes the responses from this first major finding.

Table 5.1 Separate Purchasing and Logistics Functions Can Enable Misalignment

Which of the following best describes the organizational structure for purchasing and logistics?	Percent Responding
Procurement and logistics are separate functions and are not part of a common supply chain organization	14.0%
Procurement and logistics are separate functions but are part of a common supply chain organization	45.5%
Procurement and logistics are part of the same function and are part of a common supply chain organization	12.2%
Other/not applicable	28.4%

Major Finding 2: Both Purchasing and Logistics Are Well-Aligned Independently With Their Business Unit's Strategy and Activities but Not Nearly As Well-Aligned With Each Other

The respondents provided a very strong indication that both purchasing and logistics functions are well-aligned to business unit strategy and activities. You can see this data in Table 5.2. That means both groups essentially agreed with the statements supporting the alignment of purchasing and logistics with business unit strategy (1 = strongly disagree and 5 = strongly agree). As you will see in tables 5 and 6, the issue surfaces as the alignment between purchases and logistics is not strong (not aligned with each other).

Table 5.2 Business Strategy is a Point of Purchasing and Logistics Collaboration

My Functional Area:	Purchasing	Logistics	Total Sample
Identifies opportunities to support the company's strategic direction	4.28	3.99	4.08
Understands the strategic priorities of the company's senior leadership	4.17	3.98	4.03
Adapts its strategy to the changing objectives of the company	4.21	3.89	3.99
Adapts its activities/processes to strategic changes	3.96	3.85	3.89
Maintains a common understanding with the company's senior leadership on its role in supporting strategy	3.92	3.70	3.77
Educates the company's senior leadership on the importance of procurement/logistics activities	3.72	3.63	3.66
Assesses the strategic importance of emerging trends in procurement/logistics for the company	3.60	3.51	3.54

Major Finding 3: Despite Formal Organizational Links Between Purchasing and Logistics, Interaction Between the Functions Is Typically Informal and Unstructured

Respondents were asked the level of engagement with the other function through a series of questions, where 1 = strongly disagree and 5 = strongly agree. Their answers were recorded in Table 5.3.

Of the different ways that purchasing and logistics might engage, informally working together, sharing ideas and information, and working together on a team scored the highest. More proactive approaches to collaboration, such as anticipating operational problems together and sharing resources, were by far the lowest. This supports the belief that purchasing and logistics, even when housed in the same supply chain organization, continue to operate in their own siloed worlds. Interestingly, purchasing managers perceived a much higher level of engagement.

Table 5.3 Unfortunately Purchasing and Logistics Engagement is Primarily Informal and Unstructured

My Function Engages the Other in the Following Ways:	Purchasing	Logistics	Total Sample
Informally working together	3.60	3.52	3.55
Sharing ideas and/or information	3.70	3.46	3.53
Working together as a team	3.77	3.42	3.53
Resolving operational problems together	3.75	3.38	3.49
Achieving goals collectively	3.58	3.30	3.39
Developing a mutual understanding of responsibilities	3.64	3.29	3.39
Making joint decisions about ways to improve overall operations	**3.62**	**3.16**	3.30
Anticipating operational problems together	3.32	3.12	3.18
Sharing resources	3.30	2.98	3.07

Bold values have means which are statistically different.

Major Finding 4: Maintaining Open Lines of Communication Is the Most Widely Used Technique to Foster Integration

When respondents were asked how purchasing and logistics interact, maintaining open lines of communication emerged as the most important technique. These open lines are informal and typically not systemic. Again, more proactive approaches, such as identifying potential sources of tension and establishing joint prioritization of projects, were ranked lowest (1 = strongly disagree and 5 = strongly agree). You can see the rankings of each function's priorities in Table 5.4.

We also asked respondents to indicate their functional area's performance relative to expectations, where 1 = well below expectations and 5 = well above expectations (Table 5.5). Not surprisingly, purchasing managers felt their performance relative to expectations was greatest for performance metrics over which they have the most control, such as performing to purchase price/cost objectives, supplier quality, payment terms with suppliers, and supplier responsiveness/

Table 5.4 Open Lines of Communications is Helpful but Insufficient to Drive SC Results

Purchasing/Logistics Group Tends to Work With the Other in the Following Ways:	Purchasing	Logistics	Total Sample
Maintaining open lines of communication	3.94	3.52	3.65
Combining efforts on major initiatives	3.72	3.47	3.54
Developing clear lines of managerial responsibility for implementing plans	3.38	3.24	3.28
Achieving a general level of agreement on risks/tradeoffs among projects	3.43	3.20	3.27
Coordinating project development efforts	3.53	3.16	3.27
Addressing potential sources of tension between procurement and logistics	3.21	3.14	3.16
Establishing a joint basis for prioritizing projects	3.28	2.98	3.07

No statistical differences in means.

Table 5.5 As You Might Expect, Teams Feel Better About their Own Results

My Purchasing Group's Performance Compared With Expectations for Each of the Following:	
Performing to purchase price/cost objectives	3.28
Supplier quality	3.26
Payment terms with suppliers	3.17
Supplier responsiveness/flexibility	3.11
Supplier on-time delivery	2.87
Total cost of ownership	2.83
Supplier technology contribution	2.57
Inventory investment cost for purchased goods	2.40
Transportation and logistics costs	2.40

flexibility. Performance metrics that require collaboration with logistics to achieve were all well below 3.0 on the 5-point scale.

Similarly, logistics managers felt their functional performance exceeded expectations on metrics related to customer delivery. For example, customer service level establishment, network design/network location, full, damage-free, and on-time deliveries, and inbound/outbound transportation contracting and management are all metrics that fall under their control (Table 5.6). Performance metrics that require collaboration with other areas of the supply chain (for example, forecasting accuracy, total inventory turns, reverse logistics management, and time on back-order) were among the lowest scores in the entire survey.

Finally, we conducted an analysis to determine whether the data provided any indication as to whether PLi was perceived as an important lever of overall business success (Table 5.7). While this statistic is highly subjective, the table that follows provides indications that managers from firms in the top 25% of PLi in this survey believe their firms significantly outperform their competitors as compared with managers from firms with lower PLi scores (1 = well below competitors and 5 = well above competitors). In other words, *managers believe their company achieves a significant performance premium from aligning their purchasing and logistics functions.*

Best Practices

The remainder of this chapter will report the results of a series of field interviews conducted by the University of Tennessee and affiliated faculty of major supply chain leaders such as Caterpillar, Dell,

Table 5.6 As You Might Expect, Teams Feel Less Confident in Other Team's Results

My Logistics Group's Performance Compared With Expectations for Each of the Following:	
Establishing customer service levels	3.59
Network design/network location	3.38
Full, damage-free, and on-time deliveries	3.19
Inbound/outbound transportation contracting	3.16
Inbound/outbound transportation management	3.07
Inventory planning	2.92
Logistics information-systems design and implementation	2.86
Transportation costs	2.72
Total logistics costs	2.70
Time between order receipt and delivery	2.53
Warehousing costs	2.47
Logistics performance measurement	2.43
Line-item fill rate	2.34
Inventory costs	2.18
Order fulfillment management	2.03
Finished goods inventory	2.00
Forecasting accuracy	1.98
Total inventory turns	1.93
Reverse logistics management	1.89
Time on back-order	1.63

Eastman, Ecolabs, IBM, Mondelez, and P&G. The interview results uncover best practices in purchasing and logistics integration, showing how some companies are "bending the chain."

This section also provides a helpful short list of effective leadership actions a supply chain leader can take today. The four best practices are captured in Figure 5.6.

Best Practice 1: Fully Integrated End-to-End Supply Chain Organization With Common Metrics

We have learned from decades of S&OP work that a business's *demand creation* activities are most effective when they are housed

Table 5.7 When Purchasing and Logistics Effectively Collaborate—It Drives Results

My Firm's Performance in Comparison With My Competitors	Purchasing and Logistics Alignment		
	Firms in top 25% of PLi scores	Firms in bottom 75% of PLi scores	PLi performance premium for highly aligned companies
Growth in sales	3.42°	2.91°	18%
Profit margin	3.51°	2.93°	20%
Growth in market share	3.39°	2.84°	19%
Return on investment (ROI)	3.58°	2.92°	23%
Cost reduction	3.56°	2.84°	25%

°Means are statistically different.

BEST PRACTICES

Supply chains that consistently deliver the strongest business results have the following purchasing/logistics characteristics:

1. Fully integrated end-to-end supply chain organization with common metrics

2. Talented supply chain organization that rewards people for in-depth mastery and end-to-end supply chain leadership

3. Purchasing and logistics network with an operating decision framework based on best overall total value of ownership (TVO)

4. Effective information systems and work processes that enable superior business results by providing multifunctional supply chain teams the proper tools and information

Figure 5.6 Benchmark companies drive purchasing and logistics collaboration through these best practices

in a common organization. Similarly, *demand fulfillment* activities are most effective when they are integrated under a common supply chain organization. The best, most enduring results occur when everyone in the supply chain organization is focused on delivering superb supply chain results (customer service, quality, safety, cost, cash, etc.).

For example, large, successful global consumer goods businesses have learned (the hard way) about the vital importance of fully integrated supply chain organizations. These companies are structured with an "end-to-end/fully integrated" supply chain organization headed by a common leader. These organizations include purchasing, logistics, operations/manufacturing, engineering, innovation management, quality, and others. The common supply chain leader drives an energizing vision, single direction, common scorecards, and consistent rewards. Thus, 100% of the organization is focused on meeting consumer/customer needs and delivering total value to stakeholders.

These best-in-class designs are not without challenges. Frequently, purchasing owns results beyond the supply chain, including contracts for marketing spending, indirect spending, R&D suppliers, and external contractors. This creates pressure to have an executive level purchasing manager who reports to the CEO. One global supply chain has worked through this issue by formalizing the responsibility of the purchasing VP to the global supply chain officer.

Example:

- A global chemical company has likewise leveraged a partnership between corporate purchases and the supply chain. This company found it necessary to change the language and create a culture called "integrated global supply chain" to highlight the need for purchasing and supply chain teamwork. This partnership between purchasing and the rest of the supply chain ensures a common direction and reward system.

We have found that, in organizations without a fully integrated end-to-end structure, the most effective first step is to develop these types of partnerships. The organization benefits from partners' common vision, direction, and rewards, until a more long-term structural change can be implemented.

It is important to note that these types of leadership partnerships are, by their nature, dependent on the individuals involved. It can be expected, then, that these partnerships will vary as personnel change. Therefore, it is critical that leaders view these partnerships as transitions on the path to an organizational solution.

A second challenge involves the depth of integration.

Examples:

- One Fortune 500 global supply chain leader has had an integrated end-to-end supply chain for the last three decades. The company has enjoyed improvements in cost, cash, customer service, and quality. Over the years the integration has been maintained at the top of the organization but has drifted at the category teams (middle level). Functional areas (e.g., purchasing and logistics) became convinced that, because of internal productivity improvements, they needed to become focused on their own "primary measures." Unfortunately, these middle level teams are where 90% or more of the decisions impacting cost, cash, quality, and service are made. A renewal of the original end-to-end vision at all levels of the organization is now necessary.

- A successful mid-sized company has recently implemented an integrated end-to-end supply chain design. The driving factor behind the change was the inability to deliver long term business cost goals. After a decade of strong but independent savings work by the purchasing and logistics functions, the "well was running dry." The biggest ideas were no longer inside the departments but at the supply chain integration points across the departments (e.g., optimizing piece price versus transportation cost, optimizing piece price versus sourcing location). The most systemic solution was to form and reward a fully integrated team. The organization was delivering 2.5% net savings but now, after forming an integrated team, has strong action plans to deliver the business need of 4% net cost savings.

The leadership/organizational structure is only one part of the fully integrated, end-to-end supply chain. Multi-discipline supply chain teams must be involved in strategic supplier selection and development.

You have heard the saying "Do it right the first time" your entire life. Best-in-class supply chains take this to heart. Creating the best total value supply system the first time prevents non-value added costs, quality defects, customer service defects, and unproductive inventory while most efficiently utilizing your limited resources. This is broadly accepted but difficult to execute. Day-to-day business pressures often push managers into high urgency/low value activities, diverting attention from those high value activities that can really make a lasting

Figure 5.7 Unfortunately the effort to use multi-functional teams is important but not urgent. SC leadership is key to success

impact. Figure 5.7 uses a matrix format to illustrate the relationship between urgency and value.

Best-in-class supply chains utilize multi-discipline teams to manage supplier selection and develop strategic suppliers and critical materials. This ensures the right resources are involved to develop the best end-to-end supply chain solutions. These supply chains leverage purchasing as the leader of the supplier selection/development teams. The goal is to have a clear, single point of accountability while ensuring an integrated process. These multi-discipline teams include all the relevant elements of the end-to-end supply chain (e.g., engineering, logistics, manufacturing, purchasing, innovation management, quality, six sigma resources, etc.). Moreover, best-in-class supply chains prioritize the level of resource involvement with the greatest business impact. Multi-discipline supply chain teams are heavily involved with the most important suppliers/materials while auxiliary teams manage less critical decisions. Additionally, many companies use senior, experienced (in multiple supply chain components) supply chain leaders in broader supplier selection teams with the expectation that they will resource experts when needed.

Examples:

- A global information technology leader has a simple and transparent expectation for the use of multi-discipline supplier selection and development teams. The first time an employee does not use this kind of team, he or she receives a warning. The second time results in termination. This extreme principle is being utilized to change the culture and ensure TVO requirements are delivered.

- This same supply chain leader requires that all supplier selection teams maintain responsibility for supplier development. "The development of our supplier partners is critical to delivering our long term goals. We want the accountability for selection and development to be consistent. Decisions in the selection process are owned through execution."
- A major retailer is linking merchandising with its supply chain resources on supplier selection. This same company is forming multi-discipline teams to work with private label supplier development teams. Additionally, a director of supplier collaboration has been appointed to drive faster progress in these areas.
- A major global CPG company has benefited from multi-discipline teams for multiple decades. These teams have facilitated TVO at a category or brand level. The opportunity is to multiply the scale, leveraging strong supplier partner capabilities across categories. The key action plan is to involve "other category" multi-discipline teams in their supplier selection/development processes to harvest scale within a supplier. An example of this is working to align common chemical specifications across categories to increase supplier scale/volume discounts.

Best Practice 2: Talented supply chain organization that rewards people for in-depth mastery and end-to-end supply chain leadership

For years, logistics and purchasing leaders have argued that these two vital elements of the supply chain must be in separate organizations with different recruiting, training, rewards, and rituals. Typical arguments included:

- Purchasing is an externally focused organization
- Purchasing is commercial work, not technical work
- Purchasing requires strong entrepreneurial skills
- Logistics must stay focused on delivering this week
- Logistics is busy leading inventory and customer service
- Logistics must have a strong day-to-day team relationship with manufacturing plants
- Logistics must be expert planners and APEC certified

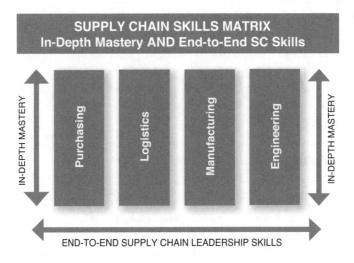

Figure 5.8 Benchmark supply chains require SC discipline technical mastery AND broad end-to-end SC experience

The most effective supply chain leaders have created new paradigms. The breakthrough improvements in cost, cash, and service lie in the seams of the supply chain, but integrated approaches are needed to achieve these benefits. Therefore, supply chains must strive for functional depth and the necessary end-to-end breadth of supply chain skills (see Figure 5.8 for a supply chain skills matrix). In our research we found multiple examples of best-in-class supply chains that have broken through to this new paradigm.

Examples:

- A large industrial equipment company is requiring its purchasing and logistics resources to come to senior business/product managers with integrated action plans and goals.

- A major global chemical company has a system to move all new supply chain managers through multiple supply chain disciplines.

- Multiple top-tier organizations are now requiring that a single senior supply chain manager with broad end-to-end skills be actively involved in upfront innovation work processes with R&D. As a result of significant corporate productivity goals, the days of sending multiple supply chain leaders are over. As one manager put it, "We must have supply chain leaders with strong purchasing and logistics skills influencing the new product/ supply chain decisions of our future."

Supply chain executive VPs are requiring that purchasing and logistics leaders do more than create "great purchasing or logistics" talent. These disciplines must develop in-depth mastery to drive results for today while simultaneously building end-to-end supply chain skills to meet the complex supply chain problems and opportunities of tomorrow. The secondary benefit of these senior leadership expectations is the creation of a larger pool of future supply chain executive leadership talent.

Best Practice 3: Purchasing and logistics network with an operating decision framework based on best overall total value of ownership (TVO = total cost of ownership plus level of customer creation)

Many purchasing teams have been using broad supplier scorecards in the supplier selection process for years.

Nevertheless, supply chains continue suffering from inaccurate prediction of supplier cost, significant quality issues/rework costs, and capacity issues because of poor supplier reliability.

Our research has shown that the existence of a supplier scorecard is insufficient to drive excellence in supplier driven supply chain metrics. Numerous supplier selection metrics must be considered. Examples can be found in Figure 5.9.

Today, most supplier selection and development is led and managed by purchasing, and decision-making is largely based on piece price. However, the most effective supply chains have successfully transitioned to decisions based on TVO. This requires broader supply chain involvement (see best practice 1) and a commitment to TVO-based decisions.

Examples:

- A world-class global information technology company uses internal supplier selection consultants to review supplier decisions. This has significantly changed the reward system. In the rapidly

SUPPLIER SELECTION METRICS	
COMPETITIVE VALUE	**LONG TERM VALUE**
Cost/Price	Innovation/Ideas
Quality	Value of Ownership
Delivery	
Reliability	**SUSTAINABLE VALUE**
Flexibility	Environmental
Responsiveness	Social/Ethical
	Compliance/Regulatory
PARTNERSHIP VALUE	
Customer Satisfaction	
Supplier Satisfaction	

Figure 5.9 Benchmark SC require broad supplier capability and metrics

changing information technology business, mistakes created by narrow piece price decisions can make or break profit goals for the company.

- A global industrial equipment company found significant defects in its supplier scorecards. The purchasing teams were measuring piece price and supplier on-time delivery. Unfortunately, 90% or more of these suppliers do not deliver the materials. This is a great example of having the wrong measures on the scorecard.

Best Practice 4: Effective information systems and work processes that enable superior business results by providing multifunctional supply chain teams the proper tools and information

Finally, a supply chain can have a fully integrated structure with talented, well-trained people who focus on total value yet still do not deliver best-in-class supplier results. Empowered teams must have the tools to execute with excellence. Robust and efficient information systems and work processes are required to support total value creation.

We have interviewed many companies that start with a holistic supplier scorecard but struggle with placing a value on customer service/

quality issues, cost of inventory, environmental incidents, reliable supply/delivery, etc.

Some of these elements have a cumulative impact. The first environmental issue may have a limited impact, but multiple incidents can cause significant legal costs, time commitment, investments and business disruption (in extreme cases). How do you place a cost on these types of elements? We have found that the best-in-class supply chains *partner with finance* to align on the value of these items. These can be intense debates. Our suggestion is to create a starting point and adjust as you learn.

Examples:

- A medium-size service-oriented company recently changed its executive vice president of supply chain. The new executive found that the reported supplier savings were not making it to the bottom line. The reward system was based on gross purchase price savings. The EVP changed the supply chain reward system, focusing the purchasing cost measure on net savings, thus creating an immediate impact on corporate results.

- Likewise, a large global consumer goods company had to change how it valued quality investments. Historically, investments for handling new materials and suppliers were approached with a "zero capital mindset." New materials/suppliers were simply brought in, and it was up to the facility managers to utilize their extremely high skilled work force to develop low-cost solutions. This had a compounding effect, in which every couple of years a major quality-based capital appropriation was required. Finance and the supply chain leader aligned on including "fair share" capital as part of these types of supplier selections to realistically model the total cost.

Robust work processes are required for managing these decisions. Clear metrics, decision authority, multi-discipline team criteria, monthly reviews, and leadership involvement are a few of these important processes.

Examples:

- A global chemical company has rigorous quarterly review processes (by chemical segmentation). The quarterly reviews include a full analysis of successes and failures, with action plans to drive continuous improvement. Additionally, the reviews are based on a holistic total value scorecard, including suppliers' work on innovation (supporting the chemical company's initiatives and internal supplier innovation).

- A global CPG executive vice president requires top supply chain leadership reviews of critical supplier decisions. "Left alone, the culture reverts to a purchasing process based on piece price. To change this culture, supply chain leaders must be actively involved in the reviews—pushing for total value, driving to determine the cost of quality/service issues, and incorporating the true cost of cash." The active leadership involvement and reviews are changing the culture, driving better decisions and training the organization on what will be rewarded.

This is complicated as the world becomes more global. Many of these processes must work across different regions. We now live in a virtual world requiring virtual processes.

Example:

- A mid-sized global service company has utilized multi-discipline teams. Because of the global nature of its business, these teams are virtual, with participants from around the world. Their inefficiencies come from the virtual work process. Teamwork is a key issue. The virtual team did not have the level of teamwork experienced in co-located facilities. The root causes were little/ no informal time for team communication, lack of teambuilding to build trust, and the time lag as the team collectively solved problems. Implementation of improved communication tools for virtual teams is a critical action plan.

Seven Actions a Supply Chain Leader Can Take Today

The value of the research, best practices, and examples is determined by how they can change your supply chain leadership. Below is a list of potential actions you could take today to make a difference in your organization and business results.

1. *Get it on business leader scorecards.* Work with your general managers/business leaders to ensure holistic measures are on

the business/general manager scorecards. Profit and cost are consistently on these high-level scorecards, but quality, cash, and customer service may not be. Including supply chain excellence measures on the business scorecard enables you to lead based on business priorities.

2. *Champion TVO.* It is not enough to talk use of total value of ownership with your direct reports. Talk the importance of total value with supplier selection and development as part of your communications (meetings, calls, printed documents, supply chain goals/action plans), participate in supplier selection and development reviews for the most strategic suppliers/materials, and ensure that the rewards for supply chain people are consistent with TVO.

3. *Make R&D your best friend.* Create a strong partnership with the research and development leader. Consider co-locating your office with the R&D leader to facilitate teamwork and symbolize a seamless technical community. The SC leader and the R&D leader should have common expectations, including active, up-front involvement in new initiative supplier decisions and product design to optimize innovation that delivers consumer, customer, supplier, community, and shareholder needs.

4. *Be clear.* Set clear expectations for use of multi-discipline teams on supplier selection. Ensure people know what process is expected for what type of suppliers. Do this publicly and in written communications. Enable your multi-discipline teams to do the work. Help your global virtual teams get the tools they need to succeed.

5. *Champion an end-to-end and integrated supply chain organization.* If your supply chain team is not end-to-end and fully integrated, create a plan to make this happen. This is not easy or straightforward leadership work in many companies. Barriers to creating your supply chain organizational vision include commercial business leaders who have other ideas, existing acquisition agreements (including personal contacts), and historical systems. Stay committed to achieving the vision, and make progress with every organizational opportunity.

Align on a common direction. If the purchasing and logistics teams have different leadership, partner with these leaders to ensure both organizations have a common supplier direction, scorecards, and rewards. This alignment can precede more

complex organizational structure changes and deliver immediate business improvement. This type of clear organizational
direction creates more leadership work, as the two leaders must
speak with a common voice. But the investment with your partner to create this common voice will reward both of you with
better decision-making (until the structural change is made).

6. *Build talent focused on the end-to-end supply chain.* Create a
principle for strong end-to-end supply chain skill requirements
for leadership positions in each supply chain discipline. Today's
business challenges require supply chain leaders who can build
strong "links" in the supply systems and resolve integration
problems. Disciplined leaders who have demonstrated successful results in multiple disciplines will strengthen the capability
of the total supply chain leadership team.

7. *Partner with finance.* Work with finance leadership to align on
how your multi-discipline teams quantify quality, customer service, environmental, sustainability, delivery, inventory, etc. A
primary leadership role is enabling the organization with clear
expectations and aligned measures. Delegation of this leadership work "freezes" most teams. Create a starting point on how
to value, learn, and adjust.

How High Is Your PLi?

Our research shows there is a tremendous benefit when firms
align purchasing and logistics activities across the value chain to facilitate collaboration.

So, how integrated are your firm's purchasing and logistics functions? *Very*, you say? Are you sure?

Why not test this with a quick check-up that will answer the question, "how high is your PLi?" Perhaps purchasing believes there is
excellent collaboration while logistics does not, or vice versa.

Send copies of Table 5.8's brief self-test to key members of your
purchasing and logistics teams, and ask them to return them to you.
See where their answers are aligned and where they are different.
This is not a scientific tool but one designed to provide insight into
how both groups view their level of collaboration. It will test the "temperature" of your PLi (Figure 5.10). This eye-opening exercise could
lead to valuable process improvements that can raise your PLi.

Table 5.8 An Effective Assessment Tool to Find Your SC Collaboration Strengths and Opportunities

How High is Your PLi?

Answer the following questions on a 1 to 5 scale.
Apply the questions based on your business.

Scale:

5 – fully implemented, producing strong results, cultural norm

3 – implemented but not a cultural norm and requires leadership reinforcement

1 – not implemented, being discussed

Question	Score	Comments
1. Do you have a fully integrated end-to-end supply chain organization where purchasing and logistics report to the same supply chain VP?		
2. Do you have one common supply chain vision, direction, and rewards system for all purchasing and logistics personnel?		
3. Do you have a common supply chain scorecard where all disciplines in the supply chain report results?		
4. Do you measure supplier selection, development and other operational decisions based on total value to your company?		
5. Does your organization have clear measures for the value of inventory, quality, and customer service to include in the total value equation?		
6. Do you utilize multi-functional teams (i.e., R&D, finance, operations, quality, engineering, logistics, purchasing) appropriate for your business to select and develop strategic suppliers and materials?		
7. Do your multi-functional purchasing and logistics teams have the information system, work process, and communication tools to do the work well?		

(*Continued*)

Table 5.8 (*Continued*)

How High is Your PLi?

Question	Score	Comments
8. Does your supply chain organization value in-depth mastery in purchasing and logistics as well as end-to-end supply chain mastery?		
9. Do the R&D and supply chain teams working jointly to create innovation that enables total value to the business?		
10. Would your business and commercial leadership (i.e., general manager, marketing VP) view the supply chain organization as fully integrated (one team) driving for best overall value for the business?		

How did you do?

40 to 50—Best in class organization benefiting from strong PLi

30 to 39—Headed in the right direction, work to do

20 to 29—Top supply chain leadership's personal involvement needed, significant work to do

10 to 19—New direction needed, significant value being lost

Transition: Bending the Chain—The Surprising Challenge of Integrating Purchasing and Logistics to Supply Chain Talent—Our Most Important Resource

By capitalizing on the opportunity to strategically integrate purchasing and logistics, companies are bending the chain to save money and optimize their supply chains. Our research has shown that financials and earnings-per-share can be positively impacted by uniting these supply chain functions. Closing the gap between purchasing and logistics is in the entire organization's best interest.

Chapter 5 put the framework in place for how companies can make this happen. However, uniting purchasing, logistics, and all

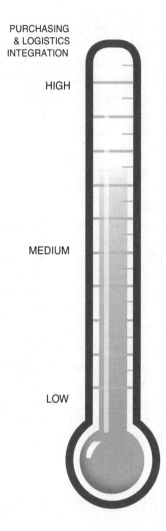

PURCHASING
& LOGISTICS
INTEGRATION

HIGH

MEDIUM

LOW

Figure 5.10 How effective is your purchasing and logistics integration (PLi)?

other supply chain functions becomes more difficult if leaders and managers are not prepared to do so. In order to bend the chain, organizations must have the proper talent in place.

It is for these reasons that our most recent publication in the game-changing white paper series is arguably the most important. *Talent* is an essential component to supply chain success. Because of the nature of supply chain management, supply chain talent is equally as dynamic.

The skills required to successfully guide an organization's supply chain are particular, and have changed drastically since the year 2000.

In some ways, supply chain talent can be compared to baseball. The most athletic players can play several field positions and still bat consistently at the plate. A team filled with only one type of player will never be as good as the well-balanced team with several utility players. The same can be said for supply chain talent. The best supply chain employees master their individual areas of responsibility and lead teams that span functions beyond their scope of responsibility. In essence, a supply chain leader must be an all-star shortstop, a serviceable outfielder, and a decent hitter.

Chapter 6 recognizes the need for quality talent. After defining the key aspects of a talent strategy, we present eight best practices for managing supply chain talent. The chapter closes with two in-depth case studies concerning truck drivers and hourly associates. Finally, as a means to help you assess your company's talent development needs, we include a supply chain skill matrix and development plan tool.

Talent is critically important to organizational success. The game-changing supply chain trends have made talent acquisition and retention even more important than it once was. By using the analyze, find, recruit, develop, and retain techniques described in this chapter, your firm will be well on its way to lasting supply chain success—success that can only be made possible with the work of extremely talented individuals and teams.

6

Supply Chain Talent—Our Most Important Resource

The Sixth in the Game-Changing Trends Series of University of Tennessee Supply Chain Management White Papers

Shay Scott, Ph.D., Mike Burnette, Paul Dittmann, Ph.D., Ted Stank, Ph.D., Chad Autry, Ph.D.

Executive Summary

Global economic growth since the 2008 to 2010 recession has created a crisis in supply chain talent management. Demand for top talent is increasing as supply chain volume and complexity rise, but the supply of that talent is decreasing. According to the Bureau of Labor and Statistics, as of February 2015, only 62.8% of the American population had a job or were actively seeking work—the lowest level since 1978. This means there are shortages in every area of the supply chain, from blue-collar laborers to senior executives. As more Baby Boomers reach retirement age, those shortages are likely only to increase.

Executives see the writing on the wall: the 2014 PricewaterhouseCoopers Global CEO survey reported 91% of CEOs recognize a need to change their strategy for attracting and recruiting talent. Amazingly, though, 61% claim they have not taken the first step to do so. The time to hesitate is through. *Your ability to ensure your supply chain talent will determine the success of your business in the coming decades.*

The labor shortage presents a problem for the global economy across the board, but the crisis is particularly pertinent to supply chain talent.

What Makes Supply Chain Talent Unique?

- Supply chain management role continues to increase. The scope and responsibility of the supply chain function has risen and supply chain teams are being asked to assume responsibility for larger swathes of the business. This has led to significant shortages for talent and the need for existing talent to have a dramatically different outlook about their roles and expectations.

- Supply chain talent must have the skills to master their own area of responsibility (typically technical in nature) and lead multifunctional processes that span the business (i.e., S&OP/IBP, program management, supplier evaluation, demand/forecasting alignment, significant capital investment decisions, etc.).

- Effective supply chain resources must be able to manage external resources and organizations that comprise modern supply chains (i.e., material suppliers, 3PLs, transportation companies, government/customs organizations, customer interfaces, etc.).

- Effective supply chain resources must understand the end-to-end, integrated supply chain. Supply chains consist of broad, complex, diverse, and technically challenging processes to master and lead. These include: suppliers, procurement, logistics/planning, manufacturing, engineering, warehousing, and transportation.

- Most supply chains are global, requiring a high level of interpersonal skills to manage, communicate, and interact with very diverse cultures. Typically these interactions utilize virtual tools that require higher level interpersonal skills.

What Is Included?

Our goal with this chapter is to make a clear case for why supply chain talent should be your highest priority—above the current quality crisis, cost compression targets, and everything else. We will share best practices from companies that have already found that focusing on talent accelerates resolution to difficult business challenges. In this paper, we will address the following:

- Key aspects of developing a talent strategy

- Results from industry research on talent management and critical recommendations based upon that research
- Eight talent management best practices gleaned from work with best-in-class industry partners
- Talent management issues in two key hourly supply chain employee categories (drivers and warehouse technicians)
- A supply chain skill matrix and development plan tool that helps assess your own supply chain talent development needs

Our Most Important Resource

In 2015, supply chain talent management is arguably the most difficult and unique of all business requirements. The skills necessary to be successful in a supply chain organization are diverse, complex, and broad. This, combined with the current market situation where the demand for most of these skills/experiences exceeds the supply, creates a huge business challenge. Our benchmark companies consistently manage talent with the following mindset: *supply chain people are truly our most important resource.* Our people are the most difficult asset to replace and the primary resource that drives supply chain results.

Applying the following approach to talent management, based off of our research and the best practices from our benchmark interviews/case studies, will help you make supply chain talent a competitive advantage for your business.

- *Talent Management*: The organizational systems and processes by which companies analyze, find, recruit, develop, and retain their supply chain human resources in support of their business strategies.
- *Diversity*: In this chapter, our discussion is based on the broadest definition of diversity to include positive leveraging of experiences, background, thought, race, culture, gender, etc.

Introduction

The importance of talent to an organization's success is something we understand early in life. Think back to those moments on the school playground when it was time to choose teams for a game of kickball, football, or whatever activity dominated your recess time. Who was chosen first? Was it the best players or those who got along well with others? Or was it those who had distinguished themselves in some way? Even as schoolchildren, we recognized the overriding importance of analyzing our talent needs and then attracting, recruiting, developing, and retaining the best fits for the tasks at hand.

As professionals and leaders, it is even more obvious that success in achieving our organizational goals is built upon having the right people in the right places. Consider the primary emphasis on talent found within professional sports franchises, medical research teams, or political coalitions. These are all groups of people striving for remarkable results, where the value of having the correct blend of skills, attitudes, and experiences cannot be understated. In fact, those groups that happen to come within striking distance of a major goal without the benefit of premium talent are usually dismissed as flukes or miracles.

The importance of managing and developing talent is certainly not a new concept to business. In the classic business book *Good to Great*, author Jim Collins emphasizes the primacy of people (talent) over direction (strategy). He does this by using a metaphor of getting the right people on the bus in the right seats, and removing the wrong people, before beginning a journey.[1] Critics of this approach point to the idea that organizations never intentionally hire inferior talent, nor do they knowingly sabotage themselves from achieving remarkable results. However, our experience interacting with hundreds of companies demonstrates that supply chain leaders often unintentionally deprioritize talent management—much to the detriment of organizational performance.

Our goal with this chapter is to make a clear case for why talent should be the highest priority—above strategy, cost compression targets, and everything else—and to share best practices from companies that have already found that talent will successfully address many other business challenges. In the following sections we will speak to:

- Key aspects of developing a talent strategy
- Results from industry research on talent management and critical recommendations based upon that research
- Eight talent management best practices gleaned from work with best-in-class industry partners

---

[1] J. Collins, *Good to Great: Why Some Companies Make the Leap—and Others Don't* (New York, NY: HarperBusiness, 2001).

Some Terms

Talent Management —The organizational systems and processes by which companies analyze, find, recruit, develop, and retain their supply chain human resources in support of their business strategies.

Supply Chain(s) —The end-to-end, integrated supply chain from the supplier's supplier to the consumer's shelf.

Supply Chain Organization —The holistic resources and teams required to deliver products to the end consumer with excellence. This includes, but is not limited to, procurement, materials management, manufacturing, engineering, process control, quality, safety/environmental, warehousing, transportation, distribution, logistics/planning (production, category, customer planning), and innovation program management. The supply chain organization is the people who support the supply chain.

Figure 6.1 Benchmark supply chains have a broad, end-to-end definition of the supply chain and its talent

- Talent management issues in two key hourly supply chain employee categories (drivers and warehouse technicians)
- A supply chain skill matrix and development plan tool that helps assess your own supply chain talent development needs

Figure 6.1 defines three important terms that will be found repeatedly in this chapter and the book as a whole.

Developing a Talent Management Strategy

Most would agree it is untenable for everyone to pursue the same top 10% of talent based upon generic factors like salary or grade point average. Yet companies appear to be doing just that. Without a clearly articulated talent strategy, human resource departments often source individual players based on the aforementioned generic rankings instead of seeking people whose skills, attitudes, and experiences make them an outstanding fit for a specific role. *The best fit for a specific role in your organization will often differ from the best fit for the same role in another organization* due to culture, priorities, and other situational factors. Emphasizing recruitment by role obtains better talent for your company and relieves some of the competition over talent. As you will read throughout this chapter, the goal of a successful talent strategy should be to harness the company's collective human resources toward an integrated effort to deliver on organizational goals.

The towers of isolation and the misaligned metrics between supply chain and human resources organizations can create dysfunction unless a company has a deliberate strategy to drive talent goals. Unless human resources teams are closely collaborating with the functional areas they serve, they become exclusively driven by metrics that do not relate to supply chain performance. Such metrics include how long a position remains open, average salary statistics, and the efficient use of training budgets across the entire organization. Furthermore, the *development* of existing talent in the firm is often deprioritized as resources are disproportionately channeled toward new hires. While not bad goals, typical human resource department priorities are often out of sync with business needs.

To be certain, talent is not just a concern for supply chain management, but one that spans the business enterprise. The 2014 PricewaterhouseCoopers Global CEO survey reported that 91% of CEOs recognize the need to change their strategies for attracting and recruiting talent, although 61% have not taken the first step to do so.[2] To add to the concern, 63% of this same group of CEOs indicated concern about the availability of key skill sets and only 34% are confident that their human resources teams are well-prepared to face this challenge. This is simply not a sustainable position.

Similar to schoolyard results, talent creates winners and losers. In fact, we believe talent will be the defining issue for supply chains over the decade to come. Supply chain management is increasingly accepted as a primary way that companies create value. In this new role, supply chain organizations are being asked to think differently and perform differently. They must have the right talent in place to do so. Buyers, plant managers, and transportation directors from the past supply chain eras do not naturally have the skill sets, attitudes, or experiences to succeed in this new world of integrated, end-to-end supply chain management. Talent development—equipping people with the necessary skills, attitudes, and experiences—must be deliberately undertaken with improved relevance to employees in their new, strategic roles.

[2] "17th Annual Global CEO Survey: Fit for the Future, Capitalizing on Global Trends," *PricewaterhouseCoopers International*, March 16, 2015, http://www.pwc.com/gx/en/ceo-survey/2014/index.jhtml

The challenges of developing a robust talent strategy are numerous. As we highlight in this chapter, shortage situations exist across all areas depending on company particulars and the supply chain involved. For example, many firms in the United States battle acute truck driver shortages. Given constraints with rail capacity and port infrastructure, the demand for truck transportation has increased and will likely continue to do so. However, the labor market continues to become less attractive to drivers. A talent and operational strategy must be developed for use in this area. If a company fails to develop this kind of strategy, the simple (yet critical) task of transporting goods to market can grind an otherwise good supply chain to a halt.

Senior supply chain leaders also recognize that there are not enough middle managers with the ability to transition an organization to an integrated end-to-end supply chain operating philosophy. How will a company internally develop this capability? Will it need to be seeded with external hires? These questions should be addressed by a comprehensive talent strategy. Issues with spot talent shortages (e.g., China, Singapore), the continued adoption of complex end-to-end supply chain philosophies, and external market forces (e.g., immigration law, shale oil boom, and cyclical construction hiring) combine to make a coherent talent strategy an absolute must in the current business environment. The days of assuming supply chain talent is available in plentiful supply are gone. We must now consider talent to be a critical resource that is in short supply. We need to engineer supply chains for talent much like the work we do for any other critical input. This will provide assurance that talented supply chain resources will be available when we need it most.

Talent Management Myths

Paralysis seems to most accurately capture the current response of many companies to the talent challenges described above. Multiple studies reveal that leaders appear flummoxed when attempting to develop a talent strategy. It is well-established that supply chain leaders tend to over-report their organizational performance and capabilities when surveyed. However, on the subject of talent, 83% said they were no better than their peers. Firms are deficient in the talent management domain, and they seem to know it.[3]

---

[3]L. Cecere, "Supply Chain Talent: The Missing Link?" *Supply Chain Insights, LLC,* November 8, 2012, http://supplychaininsights.com/supply-chain-talent-the-missing-link (retrieved January 2015); "17th Annual Global CEO Survey."

Table 6.1 Traditional Talent Myths Can Limit Your Talent Management

Supply Chain Talent Management Myths
1. Talent management is HR's responsibility.
2. Talent can neither be measured nor managed.
3. We cannot afford to spend on talent recruitment and development.
4. Talent development is primarily about teaching supply chain content.
5. A one-size-fits-all solution will work for talent development.
6. Internal (or external) resources are always better.
7. Development happens primarily in a classroom.
8. Talent development will happen naturally and informally.
9. Talent development is less important than the issue-du-jour.
10. We are so far behind that we should give up now.

Prior to presenting the results and best practices that emerged from our study on talent, it is appropriate to identify and dispel some common myths that prevent organizations from tackling the talent challenge with the priority and confidence they need to do so. We identified 10 common myths (listed in Table 6.1) that we believe impede supply chain talent development progress. In lieu of discussing each myth individually, we will present some summary assertions to guide this discussion.

The belief that talent is a fundamentally abstract topic—a step removed from financial results—undergirds many of these myths. In truth, talent acquisition and development can (and should) be firmly tied to quarterly, annual, and multi-year business results. Talent should be the foundation upon which a business strategy is built and completely integrated with it. How can a company expect to implement modern supply chain concepts (such as end-to-end, lean/six sigma, omni-channel, or global sourcing) if no one in the organization knows what they should look like? How can a supply chain team operate in global markets if they have inadequate experience there?

The Need for Ownership of the Talent Strategy

Supply chain strategy and talent strategy are heavily intertwined. A talent strategy cannot be managed without a thorough understanding of the accompanying supply chain business strategy. We believe this hits at the first of the major myths—that talent management is HR's responsibility. Supply chain leaders tend to abdicate responsibility for talent strategy often because corporate structures do not empower them to develop it on their own. However, if the human resource organization tries to develop a talent strategy without the insights of supply chain leaders, they will never grasp the rapidly evolving nature of supply chain management as compared to corporate functions such as finance or marketing. A talent strategy must be constructed through an effective partnership between supply chain and human resources leadership.

If you feel you may be behind your competition in managing supply chain talent, you need to take action and take it quickly. Early wins can be found in formalizing existing partnerships between supply chain and human resources, improving basic communication, and taking one step at a time. For example, a program developed between supply chain and human resources to identify and nurture existing key talent has the potential to strengthen the entire organization. An example is in the best practice section of this paper. It discusses how benchmark companies use "Top Talent" systems to deliver their long-term senior Supply chain executives for the company.

Talent Strategy Must Support Organizational Goals

An effective partnership between supply chain and human resources is only the first step in developing the talent needs of your business. However, human resource practices should already be designed to support organizational goals rather than departmental directives. If the supply chain has planned to cut costs, enter new markets, change its manufacturing model, improve customer service, or employ any of the dozens of other strategies, a talent strategy must be put in place to ensure that the team has the proper skills, attitude, and experience to be successful.

This seems like common sense. But 77% of organizations have no set budget or roadmap to develop supply chain talent.[4] For supply chain leaders, it is unthinkable to enter into a fiscal year without a plan for how to lower costs or improve fill rates. Yet this is exactly what we are doing with our talent! Talent acquisition and development is becoming a fiercely competitive corporate sector. A robust supply chain talent strategy can determine which company prevails.

---

[4]Cecere, "Supply Chain Talent."

The Need for Investment in Talent

We do not recommend blind investment with no clear return. Investment in talent, whether the acquisition of new talent or the development of existing resources, should carry the same stipulations as any other type of investment. Every organization decides how it will invest to build capabilities and improve performance. In most companies, this involves a lengthy list of potential capital equipment and technology enhancements that are prioritized and authorized through a leadership council of some sort. However, when it comes to talent acquisition and development, we usually do not follow the same process.

Why do we not manage talent development projects the same way we manage capital and technology projects? The processes should actually look quite similar. As part of talent development, a group of employees build some type of increased competency. This should directly translate into improved business performance. This follows the same pattern and yields the same ROI as any other project.

We need to change our approach to talent investment. By talent investment, we do not exclusively or even primarily mean competitive salaries and benefits (although they are an important part of talent strategy). Instead, we use this term to denote development expenses that raise the competencies of employees. This may include formal education programs, rotational assignments, or coaching. Talent retention is critical. Job search and training costs to replace an experienced person are two-to-four times annual salary on average. Talent investment has the potential to yield much higher returns than capital or technology improvement investments—although they are often not jointly considered.

Think About Talent Like You Think About Supply Chain Management

Supply chain leaders have the skills and experience necessary to take ownership of talent by successfully developing strategies and making wise investments. We know this because these same skills are inherent in managing supply chains. Table 6.2 illustrates how many of the concepts that are familiar to managing supply chains can be translated to talent. For example, the construction of best-of-breed solutions is often the optimal route to building organizational talent. But just as it does in software, this means that the owner must have an in-depth understanding of how each component fits into the whole.

Table 6.2 Benchmark Companies Manage Talent Like They Manage Their SC

Think about talent like you think about Supply Chain Management:
■ Have a clear vision/strategy/plan.
■ Emphasize results and tie activities to supply chain KPI's.
■ Ensure the continuity of supply including clear risk mitigation strategies for ensuring talent availability.
■ Make intelligent make/buy decisions on whether to acquire external talent or develop existing resources.
■ Continuously evolve and improve with the changing market forces.
■ Blend a mix of organizations, schools, associations, and individuals to execute a successful strategy.
■ Make talent a priority before a crisis forces you to do so.

Continuity of supply principles that are so familiar to procurement can also be applied to talent.

The challenges and opportunities for supply chain leaders in the area of talent can be overwhelming. The following section presents our research in this area and offers a five-step talent framework to divide the subject into key activities at which companies need to excel.

Talent Management Industry Research Results and Recommendations

In order to provide a broad perspective on supply chain talent development, we surveyed 126 supply chain professionals across a range of industries regarding their process for managing supply chain talent. The companies surveyed ranged in size from $400 million to $80 billion in revenue. They included retailers, manufacturers, CPG companies, and some service providers. We focused the survey on the five categories of talent management: analyze, find, recruit, develop, and retain. The Five-Step Talent Management Framework is presented in Figure 6.2. Results of the survey are reviewed in the following narrative.

Analyze

Analyze Talent: Respondents felt best about their process for documenting specific job skill requirements before launching a job search. They were a little less confident regarding their process for documenting the desired personal characteristics, or soft skills, such

Five-Step Talent Management Framework

Analyze	The work to clearly define the skills, experiences, and capabilities required in the supply chain today and for the next decade.
Find	The effort to locate a critical mass of people with the skills, experiences, and capabilities needed to deliver the supply chain goals.
Recruit	The process to attract, select, and land the resources needed.
Develop	The systems required for building skills, experiences, and capabilities in your talent to fill all the roles (at all levels) in the supply chain organization. The process to enable all people to be their best.
Retain	The systems to reinforce, support, recognize, and reward supply chain resources. The process to keep your important resources and best talent.

Figure 6.2 GSCI five-step process for effective talent management

as ability to communicate or collaborate cross-functionally. They were least confident in their process for regularly re-evaluating and adjusting skill requirements as the environment evolves. The skill set needed in yesterday's supply chain will not suffice for the supply chains of the future. See Figure 6.3 for detailed response results.

Take-away: Firms need to do a better job documenting personal skill requirements for their positions, and they need to have a process to keep all position requirements up-to-date. The sum of all these descriptions/skill requirements must enable the supply chain and the business to deliver on their goals.

Find

Find Talent: Survey respondents felt best about their ability to tap a wide range of sources in generating a pool of qualified candidates. A plethora of sources exist today, including a wide range of web tools.

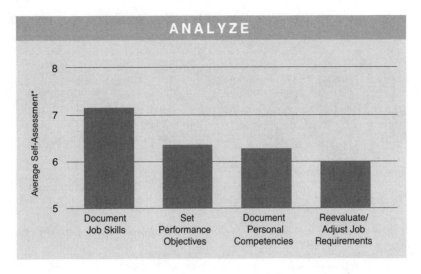

Survey Questions

• Do you thoroughly document specific job skills before launching a job search?

• Do you thoroughly document the job candidates' personal competencies (e.g., knowledge, ability to communicate, personality) before launching a job search?

• Do you develop specific performance objectives for each position before launching a job search?

• Do you regularly reevaluate and adjust competency requirements for each position as the environment evolves?

Figure 6.3 The most important step is a solid analysis, but this is frequently poorly done

Respondents were a little less enthusiastic about using these sources to develop a strong pool of candidates, and even less confident in their ability to create a strong group of diverse candidates. Many of the respondents said they work hard to maintain an up-to-date network of contacts among universities, industries, and recruiting firms. The lowest-ranked response in this section of the survey had to do with a robust process to select and evaluate executive recruiters. Many simply use hearsay to select the recruiter, and never rigorously evaluate their performance. The responses are ranked in a bar chart in Figure 6.4.

Take-Away: Generating a good pool of diverse candidates for open positions is a challenge and needs to be a strong focus in all companies. As a retail corporation's CEO put it, "Our management team needs to be diverse and essentially look and think like our range of customers." He demanded diversity—and with that emphasis, excellent results followed. In general, all parts of the process to develop a good pool of candidates need to be rigorously evaluated with continuous improvements.

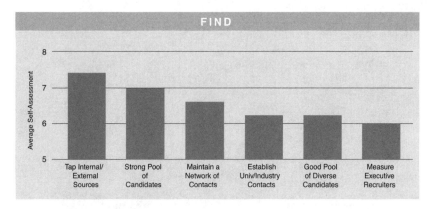

Survey Questions

• Do you tap all internal and external resources (e.g., employee referrals, social media, other web tools) to find a good slate of candidates?

• Do the recruitment processes you use identify a strong pool of highly qualified candidates?

• Do the recruitment processes you use identify a strong pool of highly diverse candidates?

• Do you spend considerable time and effort in developing a recruiting network with universities and industry contacts?

• Do you maintain an up-to-date network of contacts to tap when searching for candidates?

• Do you have a formal process to both select and evaluate executive recruiters?

Figure 6.4 We know how to find SC talent once we are clear on what we want

Recruit

Recruit Talent: As is illustrated in Figure 6.5, respondents seem to feel pretty good about hiring the people to whom they make offers, although they were less-satisfied regarding diverse candidates. The lowest-ranked score in this area involves documenting the reasons for job offer declines, with an even lower ranking for developing an action plan to address those problems. Closing the deal is critical. Otherwise, all prior work is wasted.

Take-Away: Firms need to be sure to document and analyze the reasons for job offer declines and develop an action plan to address them. Much can be learned from such an exercise and real improvements can be made.

Develop

Develop Talent: Respondents felt best about having a good process to identify top talent. Leading companies regularly identify people who can move higher in the organization. Such individuals are critical to the future success of the firm. Each company should have

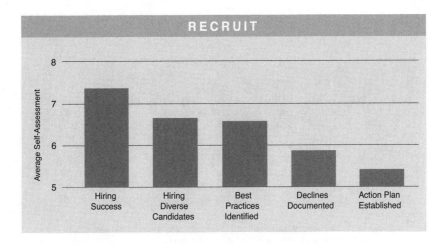

Survey Questions

• Are you successful in hiring a high percentage of job candidates to whom you make offers?

• Are you successful in hiring diverse candidates?

• Have you developed a set of best practices from successful searches in the past?

• Do you analyze and document reasons for job offer declines?

• Do you develop and implement an action plan to resolve reasons for job offer declines?

Figure 6.5 Recruiting great SC talent is a challenge as currently demand exceeds supply

a personal development plan and a coaching and mentoring process in place. Survey respondents indicated that it is reasonably common for top talent to receive mentoring and coaching. However, it does not seem to be the norm for new hires during that critical first year of adjusting to the culture of a new company.

Survey respondents also feel reasonably positive about having a good process to set individual objectives and conduct regular performance reviews. Unfortunately, this basic part of the talent development process is not ubiquitous. Some firms in our survey indicated that they had much work yet to do in this area. Most companies link pay to performance, but a substantial number feel that the pay-for-performance process is not totally fair.

Survey responses indicate that it is reasonably common for individuals in the firm to have a professional development plan, although this is far from unanimous. In-house training programs seem to be lacking for many companies.

For an in-depth look at how respondents self-assessed their talent development procedures, see Figure 6.6.

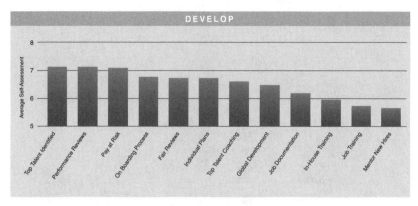

Survey Questions

• Do you have an individual development plan for each person in your organization, including appropriate certifications?

• Do you formally identify top talent (e.g., the top 10 percent) at each organization level?

• Do you make sure top talent receives special mentoring and coaching attention?

• Do you have a documented on-boarding process for new hires, including orientations, tours, cross-functional meetings, etc.?

• Do you have a mentoring program for new hires?

• Are specific objectives set for each person in the organization with regular performance reviews and feedback?

• Is variable pay (e.g., pay at risk) tied to performance?

• Do your employees respect your performance evaluation process and regard it as fair?

• Does each job have documented procedures for employees to follow?

• Does each job have a training program?

• Are there in-house training programs for areas like coaching, supervising, leadership, etc.?

• Are high potential people given opportunities to develop global expertise, including expat assignments?

Figure 6.6 Talent development is important and difficult work

Take-Away: All companies need to step up and take the time to work with each employee on a personal, professional development plan. Talent is wasted if not developed, and talent becomes obsolete at an alarming rate if not constantly challenged to grow. As one executive told us, "Our best today will not be good enough tomorrow."

Retain

Retain Talent: Contrary to supply chain leaders' assessment that they do not manage talent well, most companies feel fairly good about being able to retain top talent and diverse employees in their firms (Figure 6.7). They think they do reasonably well with the competitiveness of their compensation packages, and they also believe they do a good job at recognizing good performance. Companies indicate they are most deficient in analyzing the reasons for retention failures as well as developing action plans to correct them. Their failures to document, analyze, and develop action plans seem to be a theme throughout the survey and at each step in the talent management process.

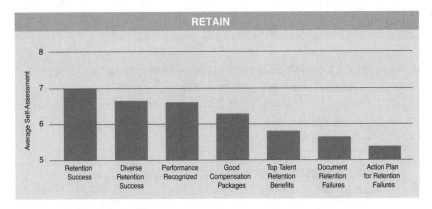

Figure 6.7 Retaining great SC talent requires strong wages/benefits and challenging work

Take-Away: Companies need to do a better job at documenting the reasons why people leave and develop actions plans to correct any problems. Replacing good talent that unexpectedly leaves is a significant expense in the short run and a major strategic problem in the long run.

Global Supply Chain Institute Talent Management Recommendations

Based on this talent survey and our previous talent research/networking, the Global Supply Chain Institute has developed three critical recommendations that should be discussed with supply chain and human resource leaders as you assess and renew your talent systems. These recommendations are summarized in Figure 6.8.

Eight Talent Management Best Practices

After completing the supply chain talent management surveys, we engaged in a series of field interviews with 11 benchmark supply chain companies to identify best practices in talent management. These companies represent a diverse set of supply chains across industries

GSCI TALENT MANAGEMENT RECOMMENDATIONS

1. **Create a clearly documented, talent development strategy** that directly supports your organization's supply chain and business strategies. This is the first and most important step. Unfortunately, we have found that creating a holistic talent strategy is often overlooked.

2. **Employ best-in-class talent development programs** that include educational and experiential components. This should include a mixture of internal and external experiences. An example could be exposing your top talent to external supply chain top talent through an educational program like the University of Tennessee's Executive MBA—Global Supply Chain (EMBA-GSC).

3. **View talent development as owned by the business and driven by ROI** instead of softer measures (manage talent like you run your supply chain). Examples include: establishing stretching talent goals, clear, hard number talent measures on your monthly scorecard, action planning teams to drive continuous improvement, annual process to renew talent systems, and linking talent to leadership reward systems.

Figure 6.8 Our research shows that effective talent strategy, talent development, and rigorous management of the process are the 3 key factors for success

such as chemicals, consumer packaged goods, food, transportation/warehousing/distribution, services, retail, and science/technology. These firms ranged in size from $15 billion to $120 billion in sales. Ten of the benchmark companies are global and one is a regional (USA) business. We selected these companies because we believe they represent best-in-class practice in their talent management systems.

We identified thirty-three different best practices that enabled these companies to analyze, find, recruit, develop, and retain the talent needed to meet their supply chain needs. These best practices are granular and specific systems. They are proven to drive results. Eight of the thirty-three best practices were widely utilized across the benchmark (versus best practices used by a minority of the benchmark companies).

In this section, we will review and discuss these eight best practices, as well as a few novel concepts that we found to be of interest.

Best Practices in Supply Chain Talent Management

1. Clear definition of the "Who"

The concept of the "who" has been reapplied by supply chain and HR professionals from marketing departments. Marketing spends significant time, effort, and money determining "who" their consumer is, "who" buys their products, "who" uses their products, and where/how their brands are found.

Benchmark supply chains in the *Analyze* element of the five-step talent framework use this same concept. One supply chain HR leader stated, "If you cannot describe who you need, anyone will do." The most important leadership work and the first step in talent management is to define the "who" in detail. From the survey data, companies struggle with soft skills on this step. Most companies are clear on technical skills, as these skills are generally covered within university supply chain management or engineering degrees. But soft skills—such as those needed to be successful in the company's culture, to do the work in unique product lines, to lead multifunctional processes, and to succeed in customer-facing roles—are frequently minimized. Leading companies document both technical and soft skill requirements in detail. They fully communicate the precise needs, and make sure they are fully aligned with the decision makers and stakeholders across the company. Some sample questions that supply chain leaders should ask themselves include:

- Are you looking for a hard worker with potential that you can develop?
- Are you looking for an experienced supply chain leader with specific expertise?
- Are you looking for a college graduate with excellent technical skills and leadership potential?
- Are you looking for a college graduate with excellent planning skills and the potential to develop technically?
- Are you looking for someone who can become a senior supply chain executive (such as a Vice President)?
- What levels of communication skills are required?

Example:

- We talked with a company that learned this lesson the hard way. They had recruited transportation and engineering graduates at a particular university for many years. These graduates were very successful in the supply chain organization of the 1980s. Since 1980, though, the supply chain organization had become global and end-to-end, expanding to include procurement, materials/production planning, manufacturing, warehousing, category planning, customer logistics, and engineering. As the years passed, the transportation students were not as effective when they were assigned roles beyond transportation. Finally, the recruiting team decided to complete an audit of the modern university course requirements and found the old transportation degrees far too narrow for the supply chain of 2015.

2. Use of Mentors, Sponsors, and First Coaches

Everyone needs help to succeed. In the *Develop* element of the five-step talent framework, having the right resources to support individual growth is a best practice. Several of the benchmark supply chains have internal data that supports these conclusions:

- New hires do significantly better with an active sponsor and/or a proven first coach
- Top talent develops faster with an active mentor or sponsor

Mentor and sponsor practices are not new. For quick reference, these roles are defined in Figure 6.9. The sponsor can be anyone in the organization. Many examples exist of technical experts effectively sponsoring talent development. This works well when the new hire needs to significantly improve skills that technical expert possesses. The sponsor can coach (by productively and directly talking about improvements, strengths, and perceptions), support, network, suggest solutions, and recommend resources.

The mentor, on the other hand, is someone in the hierarchy that owns the long-term success and development of the individual. Most companies only create formal mentors for top talent candidates, but the system could work for anyone. Mentors do not have to be a friend, but they do need to be effective role models. They must be able to

Some Terms

Sponsor —Anyone, not necessarily a "hierarchical level" dependent, in the organization who is responsible for supply chain employee coaching, advice, networking, or support. A sponsor is normally not in an employee's line organization and is <u>not</u> accountable for the person's results or career progression. A sponsor is typically utilized most frequently in a transition period (new hire, relocation, post promotion, personal crisis, etc.) but can be ongoing.

Mentor (Champion) —A supply chain leader who is accountable for an employee's success and progress in the organization. A mentor is typically two levels up in the supply chain organization. The mentor personally coaches the employee and represents the individual in supply chain personnel processes (promotion, assignment planning, top talent). Mentors are typically assigned to support top talent systems.

Note: These terms (sponsor and mentor) are not defined consistently by companies. The majority of our benchmark companies used the above definitions, but it is not unusual to work with a company that reverses the meaning of each. We have defined these to help guide you through this paper.

Figure 6.9 Sponsor/mentor work is key, but there are no common industry definitions

support and represent the person they mentor in key management systems reviews (promotion, assignment, etc.).

A first coach system is not common in most companies and is challenging to implement. This system is based on the following beliefs:

- The first manager (coach) is the most important manager in a person's career (impacts contribution, retention, etc.)
- Only about 10% to 15% of managers are good coaches

In these first coach systems, all new hires work for one of the organization's best coaches (the 10% to 15%) during their first year of employment. This ensures that new employees have a great starting coach. The downside to this system is that all new employees must start in the same areas. This can lead to disagreement in determining who the best coaches are.

3. Individual Skill and Development Plans

All companies with best-in-class supply chain talent management processes have individual skill and development plans. These plans pertain to the *Develop* element of the five-step talent framework.

The specific format and details of the tools vary by supply chain type/industry. At the end of this paper, we have included a list of principles for effective development processes.

Effective skill and development plans start with a solid definition of the skills needed to be successful in the end-to-end supply chain, in supply chain disciplines (procurement, manufacturing, warehousing, etc.), and in specific roles. This requires work by leadership to align, document, and communicate these skills. This is an ongoing effort as the supply chain and its environment are dynamic, changing the skills needed for success.

Examples:

- One of the benchmark companies focuses its individual skill and development system on leveraging the strengths of the person. They call this system "Discover Your Strengths." The company's HR manager told us, "For years we focused on improvement plans (solving weaknesses) and balanced improvement plans, but we found that if we focused on 'Discovering Your Strengths' (building on strengths) that we made more progress improving our contributions."

- We found a pilot approach in one company having a unique feedback system. The system is based on the thought that, "the best feedback is immediate." This area of the company decided to stop all feedback that is over one week old. The idea is not to ignore feedback, but to create a culture that provides and accepts immediate feedback. The initial results have been positive with this unique system as it is eliminating outdated input (not helpful to the culture) and promoting quicker resolution of differences and behavioral issues.

4. Internships/Co-Ops

All the benchmark supply chains we interviewed had internship/co-op programs. These companies feel that this system is critical to effective talent management. Internships/co-ops are in the *Recruit* element of the five-step talent framework.

Internship: This is typically a three-month role that a student accepts between the second and third or the third and fourth year of a college degree program. This could be within the same function or among two or more functions. The objectives of the internship include time for the student to learn about the values/principles/culture of the company and time for the company to determine if the student would be a successful fit within its culture. Internships are more popular than Co-Ops (described below). The internship enables the company to see more recruits in its environment and enables the students to stay on track with their college curriculums with a good summer job to support college expenses.

Co-Op: This system works like the internship except for a longer duration. The Co-Op period can be six to twelve months. The advantage for the company is that a Co-Op gives them a longer period to assess the capability and fit of the student. The advantage to the student is a similar fit assessment while providing a way to save significant money for college expenses.

Benchmark companies report hiring a 60% to 90% of all new employees from internships/Co-Ops. One supply chain leader stated, "Internships are by far the most effective system to determine if a college student will fit in your culture and have the capability to be successful."

Internships and Co-Ops Assure Fit

Fit is a talent concept that has three basic questions:

- Will the person succeed in our culture?
- Will the person be judged as a positive influence on our culture?
- Will the person help us create the culture we need/want?

This is the element of talent supply that requires the most judgment. Some key questions to think about are:

- How would you describe your current culture?
- Can you describe its current behaviors and attributes?
- What elements of your culture contribute positively to your business?
- What elements of your culture contribute negatively to your business?
- What is the culture that you want? Can you describe the behaviors you want?

These are difficult questions and may be assessed contrarily by different parts of the organization.

Fit does not mean finding people that act, look, and think like you. Supply chain professionals understand that diverse teams make better decisions. Fit is created at the sweet spot between finding someone that can excel in your culture and that can enable you to create a diverse organization (diverse thinking and experiences).

5. Top University Partnerships

Most of the benchmark companies we interviewed get the majority of their new hire supply chain talent through a partnership with a top university. The number of experienced hires for entry/middle management, technician promotions, and Internet hires is a relatively small percentage of the new talent. In the last decade, there has been much discussion in recruiting and HR journals about the efficiencies in Internet-based recruiting. All of our benchmark companies had Internet-based recruiting systems, but *none* of our benchmark companies viewed this as a productive tool.

Partnering with a university is one of the best ways to pipeline talent into the organization. Some key elements of university partnerships include:

- Prerecruiting visits to screen applicants
- Building relationships with professors by educating them on the types of students that succeed in your company and utilize their recommendations about students
- Participation in university-sponsored career fairs (including guest speaking in junior/senior classrooms)
- Active sponsorship and involvement in student professional societies and organizations
- Participation in an advisory board to impact future curriculum
- Internship/Co-Op programs jointly led by the university and company
- Utilizing past university graduates to recruit at the partner university
- Creating connections to support finding and recruiting unplanned, urgent talent needs

All of these elements are focused on the same objective: finding the students that fit your "who." The university and company partner to achieve the same goal: to place students in a role where they have the best chance for success. University partnerships are in the *Find* and *Recruit* elements of the five-step talent framework.

Example:

- One benchmark company has a unique approach to the recruitment and development process: they form an "end-to-end" talent team. The team is made up of line supply chain managers and HR resources. This team recruits, onboards, trains, and develops new supply chain resources. This system reapplies a key supply chain management principle—the end-to-end system—to talent. The talent team and new hires experience the same results improvements at transition points by avoiding handoffs between the processes.

6. Top-Talent Systems

Best-in-class supply chain organizations have a separate, ongoing top-talent system. The intent of the system is to identify current leadership talent potential and ensure that this top talent is prepared to become a great executive supply chain leader. It is consistent with wanting everyone to succeed and recognizes that a path of average coaches, assignment timelines, projects, and experiences will not develop executive supply chain talent fast enough. Key components of best-in-class systems include:

- A clear, diverse list of top talent (with aligned attributes, at middle and top levels of the organization) established by demonstrated leadership behavior. This list needs to be aligned with the key stakeholders in executive supply chain decisions.

- A system to manage assignment and career plans (these will be shorter assignments, high-exposure roles, and tougher assignments, and they should include significant commercial interface). This needs to be an ongoing system to ensure it can respond to the business needs. Our benchmark companies' formal review frequencies range from every six to every twelve months.

- An active mentoring system that provides a second personal coach and advocate who are external to the function.

- A formal leadership exposure system (with a process owner) to plan interactions between the top talent and supply chain/corporate leadership, so that all supply chain leaders know the top talent personally.

Top talent systems are included in the *Develop* and *Retain* elements of the five-step talent framework.

Example:

One of the benchmark company's supply chain executives has a personal story on top talent system. He was hired into another function (IT) within his company. He progressed well in his initial assignments and was identified as a potential, future corporate executive. He was selected for the company top talent system and assigned a mentor. As a part of his work with the mentor, they found an opportunity for a developmental assignment in another function (supply chain) to better leverage his interests and skills. This assignment is helping him create a broader impact on business results. This top talent system is working well to provide the broad range of experiences/assignments to build its corporate leaders. His current broadening role as a supply chain VP is a tangible example of how this leadership development process is working.

7. Hire for Overall Supply Chain— Not for a Specific Job

The large, global supply chain benchmark companies that we interviewed hire for overall supply chain needs—not for specific jobs. This best practice is likely more relevant to companies large enough to have staffing levels that can support it. As we described in the previous chapter, effective organizations have a balance of technical masters, leaders with experience in a supply chain discipline, employees with strong end-to-end supply chains skills, and people who can see the overall integrated supply system and solve problems in a way that creates the highest total value to the company. A supply chain with this skill design requires finding talent that is capable of success in multiple disciplines—such as planning, procurement, manufacturing, engineering, warehousing, demand management, etc.

Hiring new employees for the supply chain is not only good for the company but also the student. Most students want to join a company for a career, not a job opening. They want to see themselves as a strong, contributing part of the team long-term.

Example:

- One of the global CPG companies we interviewed feels that supply chain is one of the most difficult functions to recruit.

They recruit for strong technical skills in the College of Engineering. They recruit for the strong logistics/procurement skills in the College of Business (supply chain management). This leaves the company with the requirement to train engineers on logistics/procurement, and to train the supply chain students on technical skills. Ideally, the company would like to find students with all the skills needed to succeed in the supply chain. These students would have a solid combination of technical, problem solving, and analytical skills as well as strong commercial skills in business management, planning, and leadership.

8. Active Diversity Program

The final best practice from the benchmark interviews is an important element of the organization's culture: diversity. This is a critical part of all five elements of the talent framework. A strong plan to create the diverse culture must be a key component of each of the seven best practices mentioned above. Benchmark companies:

- Include diversity in everything they do. It is a part of the fabric of the organization. It is overtly visible in the talent framework (analyze, find, recruit, develop, and retain).
- Create systems to support the retention of their diverse talent. The benchmark companies have a wide range of these support systems that include dual-career, flexible holidays (religious preferences), on-site day care, company support networks, etc.
- Create an environment for diversity of thought across levels, roles, functions, and partners in the organization.

Example:

- One of the most popular and widely used best practices focuses on dual-career spouses. We had several examples from the benchmark companies where the dual-career system was responsible for retaining top supply chain talent. One example involved a couple where the spouse had a career with a different company. The spouse was transferred to Singapore from Europe. After reviewing the situation, the company decided to transfer its employee—the one who been transferred from Europe— to Singapore as well. This meant creating a new role for the employee in order to retain them. The subsequent work completed in Singapore was outstanding, with a significant return on the transfer investment. In the end, it was a win/win/win. The company retained the employee, and the broader organization saw first-hand the positive support for dual-career couples.

Final Thought on Best Practices

These eight best practices come from years of experience in talent management by these benchmark companies. The landscape of supplying talent for your supply chain is dynamic. Supply chains are exponentially more complex because of globalization, regulation, competition, and countless other factors. The Millennial generation of students has new/different needs. The best supply chains will continue to evolve to match these changes. One of our benchmark companies is addressing this macro challenge by "managing talent just like we manage the supply chain. We understand our business, understand the competition, set goals, create action plans, scorecard our results, analyze the data, celebrate our successes, and continuously improve." This type of rigor is a winning approach for this company.

The Eight Supply Chain Talent Management Best Practices are summarized in Figure 6.10.

BEST PRACTICES IN SUPPLY CHAIN TALENT MANAGEMENT

1. Clear Definition of the "Who"

2. Use of Mentors, Sponsors, and First Coaches

3. Individual Skill and Development Plans

4. Internships/Co-Ops

5. Top University Partnerships

6. Top Talent Systems

7. Hire for Overall Supply Chain (Not a Specific Job)

8. Active Diversity Program

Figure 6.10 Benchmark SC companies follow these best practices

Case Studies in Hourly Supply Chain Talent Management

The journey to develop a holistic talent strategy is complex. It requires leadership with a "constant purpose" and it requires looking at all levels in the organization. Hourly associates are equally important in running a successful world-class supply chain organization. We have included two case studies (driver talent and technician talent) as examples in the work to create a holistic talent strategy that focuses not only on *managerial* talent issues, but on hourly employee categories as well.

Case Study 1: The Great Driver Shortage

In the past decades, many had predicted a shortage in truck drivers by the mid-to-late 2000s. It is no longer a prediction: it is here. The Great Recession delayed the onset of the shortage and provided a false sense of security, but now the shortage is a real problem. Some estimate the shortage today is about 40,000–50,000 and growing rapidly. Another estimate has the shortage increasing to over 300,000 drivers before it peaks in 10 years. Those numbers could be catastrophic. Such a shortfall would amount to a 20% gap between demand and supply. One estimate is that only 25,000 new drivers are being added annually, not nearly enough to keep up with the number that are needed.

The driver shortage is already creating enormous turnover. Today, driver turnover hovers well over 100% for the average company employing long haul drivers. There are numerous reports of trucking companies refusing hundreds of loads every day, resulting in a major revenue loss for these companies.

Why the Shortage

Why was this shortage so easy to predict, and, most importantly, how long will it continue? Some factors that led to the great driver shortage are:

Demographics

Baby boomers who have been in the industry for twenty-five years or more are retiring, and younger workers have different lifestyle expectations. Only 6% of drivers are age thirty-five or younger. The average driver is fifty to fifty-five years old.

Competition from other industries

During the Great Recession, many drivers were forced to leave the industry. Today, many of them work instead in the recovering construction, oil, and gas industries.

Lifestyle

Fewer people want to be away from home for a week or two at a time, living is the back of truck—especially with the pay as low as it is. One driver said, "The shortage is the direct result of bad pay and a terrible lifestyle. If your wheels are not turning, you are not making the pathetically low $0.40/mile we are paid. And it is terrible for health. Try living and sleeping in a truck, eating at fast food joints every day, and changing your sleep schedule based on your driving schedule. Try being home for just twenty-four hours every two weeks." Another driver pointed to the fact that "driving an eighteen wheeler is very hard work: you cannot take medication, you are constantly monitored, and you have no social life. All that for $40,000 a year."

Wage rates

Driver wages did not rise as fast as wage rates in general over the 2000–2013 period. The current $40,000 to 45,000 pay rate lags behind overall wage inflation in the economy. On an inflation-adjusted basis, one estimate shows that drivers make 6% to 8% less in real terms than they did twenty-five years ago, in 1990. Thirty years ago, the average truck driver earned four times the wage of a food service worker. Today the $41,000 average wage is only 1.8 times higher. During much of this time, the margins of trucking companies were constantly squeezed, making significant wage increases impossible. In addition, hours of service (HOS) rules limit the number of hours a driver can work, which in turn depresses their income since truck drivers are often paid per mile driven. But truck driver wages are now going up. We have read about recent pay increases between 10% and 15%.

Big brother

People who enter the profession today know they are being watched every second they are on the road. An "eye in the sky" is constantly monitoring their location, speed, and many other variables. Many companies even have cameras in the trucks themselves.

Stricter driver requirements

Drivers have to be twenty-one years old to obtain a commercial driver's license (CDL), creating a major gap between high school graduation and the time they can enter the profession. For liability and practical reasons, companies must consider work history, driving record, and any legal problems. One estimate is that compliance, safety, and accountability (CSA) requirements reduced the driver pool by 5% to 7%.

Best Practices for Retaining/Attracting Drivers

Many enterprises are developing a creative list of ideas to attract and retain drivers in addition to the obvious approach of simply offering more money. Below is a list of such ideas. You can not do everything, so you will need to prioritize and develop a multi-year plan to address this problem. We have grouped these ideas into the *Talent Management Framework* discussed earlier in the Chapter.

Find

Find drivers in high unemployment areas and offer to relocate them to where they are needed. This goes together with another trend, namely a substantial increase in the number of driver recruiters that companies employ.

Expand the driver pool. Look to aggressively identify immigrants, veterans, retirees, and women.

- *Female drivers* are a major opportunity because they currently represent just over 5% of the total number of drivers. They have proven to do the job very well and very safely. The Women-in-Trucking organization was formed in 2007 to address the obstacles that prevent women from entering the profession. Company cultures must value and respect women. Truck stops need to develop environments friendly to women.

- Many trucking firms are successful hiring returning *veterans*. They are on the job market. Not only is this clearly it is the right thing to do, the skills developed in the military match the skills needed for strong drivers.

- President Obama announced recently that he is granting temporary legal status and work permits to nearly 5 million illegal *immigrants*. This group, along with legal immigrants, may represent a huge pool of potential drivers. The truck driver position could be a pathway toward permanent status. This effort may even involve language training. In fact, it may be critical to have bilingual supervisors.

Work toward lowering the legal age to get a license. The gap between eighteen and twenty-one is significant. The age gap is enough time for people to become established in another field. Reducing the driver age requirement is somewhat controversial, with pros and cons; but many eighteen-year-olds are ready to enter this occupation.

Recruit

Help with retirement. Provide a 401(k) savings program with company match. One company matches the employee's contributions up to 6% of pay.

Offer creative benefits, such as the following:

- At one trucking company, driver associates are admitted into a club after five years of service. This club gives them the opportunity to attend a variety of sporting events, such as NFL games, NASCAR races, dinners, and many other unique opportunities all paid for by the company (including travel, hotel, and food).
- Many trucking companies provide military leave. Drivers have guaranteed time off for military training and service, with no vacation time required. Some firms offer tuition reimbursement. After the driver goes to driving school and gets the CDL, one company will help foot the bill for recent driving school graduates.
- Some companies provide more paid vacation. That can look like:
  o 1 year of service = 1 week paid vacation
  o 2 years of service = 2 weeks paid vacation
  o 7 years of service = 3 weeks paid vacation
  o 15 years of service = 4 weeks paid vacation
- Of course there are some options to provide more cash to drivers beyond simply increasing base pay. Bonus and/or incentive plans and signing bonuses are examples of this.

Develop

Have high-quality training and on-boarding processes. We have heard from the industry that a majority of drivers say they feel no allegiance to the company they work for. Building loyalty begins the first day of employment with the on-boarding process.

Form a relationship with accredited driver schools, and pick up the cost of licensing (up to $6,000) for drivers who commit to stay for a certain length of time. This would be a great area of focus for more community colleges. Some companies have had success with job fairs at driving schools.

Retain

Treat drivers with respect. This goes a long way. Everyone wants recognition and appreciation. As one executive said, drivers join companies but leave supervisors. Often they leave in the first 180 days. Therefore, front-line management needs to be re-educated regarding how to treat drivers. In fact, according to the surveys, respect is more important than money for many people. As one driver said, "At my current employer, when I arrive it is 'Hello Rick, how are you doing?' At my last company it was 'Hello 81976.'" It is important to talk to drivers and ask their opinion on improvements. After all, they know a lot more about the job than management does.

Create clear, predictable work schedules. Many drivers highly value stability and predictability.

Create flexible work arrangements. Customize jobs based on the driver's preferred lifestyle, and let drivers participate in the design of their work schedules.

Develop a career path with progressive pay raises available based on longevity. Offer opportunities to those drivers who are inclined to move into dispatch and management.

Provide better equipment, and more comfortable cabs for drivers. Spec equipment with women and younger drivers in mind (e.g., determine if automatic transmission would enable a wider pool of drivers).

Find a way to get drivers home more often. Operations research scientists and mathematical modelers can develop optimal schedules for drivers. Many drivers tell us that getting home is a major factor in getting them to stay with their company. Drivers who have a local route allowing them to be home every day have an approximately 30% lower turnover rate. That is less than one fourth of over-the-road, long haul drivers. Also, it is important to make sure that drivers have a convenient way to communicate with their families when they are on the road.

Collaborate. Shippers and their transportation providers should work together to identify ways that shipments can be optimized to allow drivers to have single-day routes. They should also find ways to reduce the total number of shipments so a smaller number of drivers are required across the network.

Eliminate the driver! This may sound radical but is potentially more realistic than we might think—and within the lifetimes of some reading this paper. A huge amount of technology is advancing rapidly in this area. Trucks may lead the way in the coming driverless revolution, perhaps someday beginning with caravans on the open road, with one human monitoring several rigs simultaneously.

Recommendation: Solving The Great Driver Shortage

All companies that employ drivers directly or indirectly need to develop a formal plan to address the great driver shortage. This is a problem for both shippers and their carriers, and ideally they would work together on solutions. The shortage is not going away any time soon, and it cries out for a proactive strategy. This strategy needs to be developed by talking with truckers and benchmarking best-in-class solutions that others are employing. It will be important to develop a prioritized action plan. Make an assessment after completion of a critical few action items. Early wins will be critical before proceeding down the list of priorities. It is important develop a focused and effective plan. This will likely be a multi-year implementation.

When we look back on this era, we may conclude that the best companies dealt with the great driver shortage with a multi-year strategy grounded in benchmarking and driver feedback. The "losers," so to speak, will be those who pursued the "idea of the month" without any cohesive strategy.

We recommend the six-step approach in Figure 6.11 when it comes to addressing the driver shortage:

Case Study 2: Supply Chain Technician and Warehouse Personnel Challenges

All of the focus in the media seems to be on the great driver shortage. But there are supply chain talent shortages across virtually every hourly associate resource group. One serious shortage is brewing with supply chain technicians and mechanics. With the improvements in the United States and global economies, and in combination with

1. Benchmark best-in-class companies.

2. Talk to drivers.

3. With a team, assemble a list of all possible ideas to address the problem. Use the above list as a thought starter.

4. For each idea, estimate its potential net return and its ease of implementation.

5. Prioritize ideas starting with quick win opportunities, high net return, and a straightforward implementation.

6. Lay out a project plan over 2–4 years, select a project manager, and implement using a disciplined project and change management approach.

Figure 6.11 The driver shortage is real—benchmark companies (retention) drive these capabilities

the recent reshoring shift, the demand for these roles is increasing. Nationally, US trade schools are not producing enough mechanics and technicians with the requisite computer skills needed to tackle complex supply chain issues. Demand is exceeding supply.

We interviewed executives from ProLogistix, a leading staffing supplier for warehouse and distribution jobs. ProLogistix has compiled employee survey data since 2010. The data are interesting and signal the following potential conclusions:

- The supply chain hourly associate environment is significantly different today versus the way it was in 2008. We have shifted from a surplus of such supply chain skills in a down economy to a shortage of supply chain skills in a recovering economy.

- Supply chain technician wages have been basically flat during the 2002 to 2012 period while the CPI is up, creating a 15% to 20% wage gap.

- Recent survey data on employee needs has shifted from "fortunate to have a job" to "my job does not have enough benefits."

Table 6.3 shows how dramatic this shift has been over the past seven years.

Table 6.3 As the Economy Improves Technician Needs Change

Area of Concern of Hourly Associates	Supply Greater Than Demand (2008)	Demand Greater Than Supply (2015)
Shift (working hours)	Any shift	1st shift or flex to personal need
Attendance	Lower	Higher
Pay	#1 – Most Important	#1 – Most Important
Job security	#2	#2
Paid time off	Lower	Higher
Advancement opportunity	Lower	Higher
Switch for better job	Lower	Higher
Shift length (8,10,12)	Any	8 or 10 hour strongly preferred

This shift in workplace dynamics is forcing companies to initiate changes, including:

- *Adjusting wages*—higher salaries entice employees to stay with companies.
- *Offering improved benefits*—specifically with flexible shift schedules, time off with pay, new rewards/recognition systems, and "stay bonuses."
- *Sponsoring technical schools and scholarship/aid to the best students*—technical schools are a key source of supply chain technician talent.
- *Providing internal training programs*—technical schools are not keeping up with demand. Other sources of talent supply come from experienced workers in the military, on farms, and in automobile mechanic shops—talent that will need additional training.

The bottom line is that companies need a strategy to deal with the growing shortage of mechanics and technicians just as they do for drivers.

Summary

In 2015, supply chain talent management is extremely challenging work. Supply chain talent is arguably the most difficult and unique of all the business requirements. The skills required to be successful in a supply chain organization are diverse, complex, and broad. This combined with the current market situation where the demand for most of these skills and experiences exceeds the supply creates a huge business challenge. The shortages are across the board in every supply chain skill set and include extreme shortages in drivers, technicians, mechanics, and first/mid-level managers.

Our benchmark companies consistently manage talent with the following mindset: *supply chain people are truly our most important resource*. People are the most difficult asset to replace and they are the primary resource that drives supply chain results.

The five-step talent model will help you think through your business challenges. Start with step 1, *Analyze*. It is the first and most important step, but frequently the step with the least quality work. Being clear on the skills, experience, and capabilities for the talent you acquire is the foundation. Make sure your talent will deliver what your business and your supply chain need today and what they will require for the next decade.

We have detailed best practices from some of the most elite global supply chain organizations. These best practices have been tested and proven over the years in complex, diverse, and global supply chains. Use this opportunity to renew your supply chain talent strategy and systems to create the talent and culture that you need in the intensely competitive global environment. Applying the learning from our research and the best practices from our benchmark interviews and case studies will help enable you make supply chain talent a competitive advantage for your business.

Finally, most supply chain leaders are familiar with the following, simple priority setting model in Figure 6.12. Unfortunately, talent management normally falls in the "important, but not urgent" section—the most difficult priority to manage. Remember—effective supply chain leaders make supply chain talent a "High Urgency/High Importance" priority!

Supply Chain Talent Skill Matrix and Development Plan Tool

During the research and interviews for this talent management paper, we found that the most frequent request from supply chain leaders—including the benchmark supply chain firms—was for examples of talent skill development tools. This is a complex request, as these tools would need to vary based on the types of supply chains

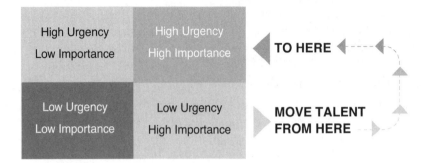

High Urgency	High Urgency	
Low Importance	High Importance	**TO HERE**
Low Urgency	Low Urgency	**MOVE TALENT**
Low Importance	High Importance	**FROM HERE**

Figure 6.12 Unfortunately talent work is important but not urgent requiring strong SC leadership

you lead (heavy equipment, automotive, air, medical supplies, pharmaceutical, CPG, service, etc.).

There are a few skill development tool principles:

- *Keep it simple.* Development tools only work if the individual uses them. The tools need to be 100% value-added to the people involved. Most development plans fail from their own weight.
- *Focus on the key elements.* Many skill tools are too complex and difficult to keep current. Remember that job descriptions should have the detail. The development tools only need enough detail to facilitate great personal action plans. Development tool key elements can vary by supply chain type but frequently include:
  o Skill development plan
  o Career and assignment plan
  o 360 degree feedback (summary)
  o Personal Annual Action Plan (what I plan to do this year based on this skill, career, and feedback information)
- *Ongoing process.* This tool cannot reside in a file cabinet. You need to review/update in regular one-on-ones with your manager. Additionally, an annual review with the next level manager, sponsors, and mentors is normally judged as helpful.

Finally, in Table 6.4, we have included a partial grid of a skill matrix. We strongly recommend that your organization complete the work to clarify the skills you need in key roles to deliver great supply chain results. This exercise is important in high-performing supply chain organizations. It enables your talent to focus on the right skills to develop. It also provides your people with a framework to create and execute a skill development plan.

Table 6.4 Effective Talent Development Plans are Most Effective When Companies Are Clear on Skill Requirements

Skill Development Plan   **Name:** _____

SC Discipline	Skills Area (Examples)	Provided by SC Discipline				Current Assessment	Individual Plan — Skill plan (education, experiences, projects, role models, etc.)
		Basic	Intermediate	High	Master		
Planning	**Production Planning Process**	Flow chart system; Manage day to day issues; Maintenance of systems	Lead group planning meetings; Train new employees; Can resolve 95% of daily issues	Set up process new products; Can solve 99.5% of daily issues; Lead improvement projects	Internal and external resource; Creates the new system improvements; Trains the trainers	Intermediate	1. Accept short term project with key suppliers/manufacturing on production plan process improvements 2. Attend CSCMP online seminar on Benchmark Production Planning systems
	Supplier Reliability	Understands key suppliers; Can evaluate supplier reliability results; Adjust production plans to reflect supplier demonstrated performance	Can resolve 95% of daily issues; Effective at working projects with suppliers to improve reliability; Train new employees	Set up process for new materials; Lead process for supplier reliability improvement; Solves most complex problems	Internal and external resource; Creates the new system improvements; Trains the trainers	Intermediate	1. Lead all suppliers' elements of planning for the new, major product restage this year

Table 6.4 (*Continued*)

Skill Development Plan						Name: _____	
		Provided by SC Discipline					Individual Plan
SC Discipline	Skills Area (Examples)	Basic	Intermediate	High	Master	Current Assessment	Skill plan (education, experiences, projects, role models, etc.)
Manufacture Production Capability		Create/communicate production plan Can flow chart manufacturing systems Qualified as basic equipment operator	Teach production teams importance of being highly reliable to results Train new employees	Set up process for new equipment Member of mfg leadership team Lead improvement projects	Internal and external resource Creates the new, system improvements Trains the trainers	Intermediate	1. Consider a career developmental assignment in manufacturing—gather data
Information System Skills		Utilize IT systems to create production plan Generate all reports from planning system Can problem solve 70% of daily issues	Can support IT's problem solving if planning system issues Train new employees	Validate IT systems for all changes with the IT resources Work joint improvement projects	Internal and external resource Creates the new system improvements Trains the trainers	Intermediate	1. Become member of multifunctional team to implement latest SAP system upgrade 2. Attend corporate training on "Getting the most from our SAP systems"

Table 6.4 (*Continued*)

Skill Development Plan					Name: _____	
	Skills Area	Provided by SC Discipline				Individual Plan
SC Discipline	(Examples)	Basic	Intermediate	High	Master	Current Assessment	Skill plan (education, experiences, projects, role models, etc.)
Inventory Parameters (master data tables)		Understands planning parameters Able read/understand master data tables Can develop and maintain bill of materials	Sets parameters based on demonstrated performance Can lead monthly data table review Provides basic training	Owns planning parameters Sponsors action plans to improve supplier and mfg capability Represent SC on corporate team	Internal and external resource Creates the new system improvements Trains the trainers Able to lead corporate team if needed	Basic	1. Accept responsibility to be the department system owner for monthly master data review
Customer Service and Inventory Results Management		Calculate monthly scorecard data Active involvement in action planning Present data/plans to mid level supply chain mangement	Leads the monthly results review Leads the action planning process Develops annual improvement plan	Able to review service and inventory results with multi-functional leaders Works jointly - multi-functional teams	Internal and external resource Creates the new system improvements Trains trainers	Intermediate	

Conclusion

What does shifting supply chain landscape mean for you and your company? We believe that these ten areas will need to dealt with immediately if you want your supply chain to be the world-class source of competitive advantage your firm so desperately needs. You will have to prioritize and address the most pressing items first. Get some early wins before proceeding down your list of priorities. Do not try to do too much at once. Focus and complete; do not launch too many projects and spin your wheels.

The contents of these chapters may be overwhelming at first. When it comes to addressing game-changing supply chain trends, how do you know where to start? To help you prioritize, we suggest that your leaders, your teams, and your cross-functional partners assess your supply chain using the questions below. Your scoring results should form the basis for a consensus discussion to set your priorities. With this book as your guide, all that is left is to get started in the highly challenging—and extremely rewarding—journey to *change your supply chain game*.

Table 1 To change your supply chain game, supply chain managers must grasp each of the topics covered in this book. Once they have done so they will be able to effectively assess their supply chains based on the criteria contained within. This final assessment is one way of rating just how much your supply chain is keeping up with the changing game.

Please rate each item on a 1–10 scale, with 10 being world-class	1–10
Supply Chain Strategy	
1. Do you have a documented, multi-year supply chain strategy?	
2. Do you understand the current and future needs of your customers?	
3. Have you identified the supply chain mega trends? Do you have a plan to deal with them?	
4. Do you understand what is happening in your competitors' supply chains?	

Table 1 *(Continued)*

Please rate each item on a 1–10 scale, with 10 being world-class	1–10
Talent	
5. Do you have a documented, professional development plan for all employees?	
6. Do you generate a strong pool of diverse candidates for open positions? Are you successful in hiring them?	
7. Do you understand the reasons talented people leave your organization? Do you have a plan to keep this from happening?	
8. Do you have a talent needs plan with documented skill requirements for the future?	
Internal Collaboration	
9. Do you have a mature and effective S&OP (or demand/supply integration) process?	
10. Do you have good cooperation and collaboration between the purchasing and logistics?	
11. Do you have excellent cross-functional teamwork and strategy buy-in?	
External Collaboration	
12. Do you segment customers and suppliers? Do you focus collaboration efforts on the most critical customers and suppliers?	
13. Do you measure results jointly with your partner firms?	
14. Do you share benefits to create a win-win environment?	
15. Can you identify real improvements in the business due to collaboration efforts in cost, inventory, and customer service?	
Customer and External Focus	
16. Do you understand the current and future supply chain needs of your customers?	
17. Do you benchmark aggressively so that you have a good understanding of best in class?	
Technology	
18. Do you understand the new technology impacting the supply chain field?	
19. Do you know what questions to ask and how to manage successful technology implementations?	
Execution	
20. Are your supply chain projects delivered on time, on budget, and on the promised benefit?	
21. Do you put high priority on change management to ensure the full buy-in of all of the stakeholders impacted by supply chain projects?	

Table 1 *(Continued)*

Please rate each item on a 1–10 scale, with 10 being world-class	1–10
Risk	
22. Do you have a good process to identify all of the risks facing your supply chain? 23. Do you prioritize supply chain risks and do you have a plan to manage/mitigate the most critical risks?	
Global	
24. Do you have a robust process to make global sourcing decisions that is strategy-driven, fact-based, and supported by a total cost of ownership? 25. Do you have strong global supplier partnerships? 26. Do you have clear visibility across your global supply chain? Do you have a rapid response capability?	
Metrics	
27. Do you have a logical metrics framework that is tied to world-class goals? Is this driving the right behaviors?	
Total Points	

• 0–100 points: Your supply chain operation is deficient. You should benchmark best-in-class operations and develop.

• 101–150 points: You have an average supply chain. You are in a solid position on which to build. You now need a multi-year plan to improve your supply chain processes and performance.

• 151–200 points: You have a good-to-excellent supply chain. Honestly assess and address your weaknesses, and continuously build on your many strengths.

• 201+ points: You have an outstanding supply chain that is approaching world-class quality. Make sure you keep it that way. You and your competitors are raising the bar everyday. Keep challenging yourself to remain at the top.

Sponsor Table The Following Sponsors Played an Integral Role in the
Composition of These White Papers

SECTION	CORPORATE SPONSOR
Game-Changing Trends in Supply Chain	**EY ERNST & YOUNG** Quality In Everything We Do **TERRA** TECHNOLOGY
Global Supply Chains	**BT**
Managing Risk in the Global Supply Chain	ups **UPS Capital**
The ABCs of DCs - Distribution Center Management: A Best Practices Overview	**KENCO**
Bending the Chain - The Surprising Challenge of Integrating Purchasing and Logistics	IBM
Supply Chain Talent - Our Most Important Resource	**Ryder** Ever better.

Index

Note: Page numbers followed by "*f*," "*n*," and "*t*" indicate figures, notes, and tables respectively.